Lasse Landt

Optical and Electronic Properties of Diamondoids

Lasse Landt

Optical and Electronic Properties of Diamondoids

Experiments on the size and shape dependence of the finite size effects in ideal nanodiamond crystals

Südwestdeutscher Verlag für Hochschulschriften

Imprint
Any brand names and product names mentioned in this book are subject to trademark, brand or patent protection and are trademarks or registered trademarks of their respective holders. The use of brand names, product names, common names, trade names, product descriptions etc. even without a particular marking in this work is in no way to be construed to mean that such names may be regarded as unrestricted in respect of trademark and brand protection legislation and could thus be used by anyone.

Publisher:
Südwestdeutscher Verlag für Hochschulschriften
is a trademark of
Dodo Books Indian Ocean Ltd., member of the OmniScriptum S.R.L Publishing group
str. A.Russo 15, of. 61, Chisinau-2068, Republic of Moldova Europe
Printed at: see last page
ISBN: 978-3-8381-2631-9

Copyright © Lasse Landt
Copyright © 2011 Dodo Books Indian Ocean Ltd., member of the OmniScriptum S.R.L Publishing group

Abstract

In this work the optical properties of diamondoids, a new form of perfectly size- and shape-selected, neutral, and hydrogen-passivated diamond nanocrystals, are investigated. The absorption and luminescence properties are studied as a function of size and shape and the optical gap of the investigated diamondoid species has been determined. The shape is found to dominate the optical response of the diamondoid outweighing size effects in the investigated size range. According to their growth scheme and their absorption behavior the diamondoids are categorized as 1D, 2D and 3D nanodiamond structures. The tetrahedral $C_{26}H_{32}$ cluster is identified as the smallest diamond nanostructure to exhibit bulk-like absorption behavior. Further, diamondoids are shown to exhibit photoluminescence in the ultraviolet spectral region. The spectra for eight diamondoids of different sizes and shapes have been recorded. The photoluminescence is spectrally broad and only little size-dependent. A spectral structure is observed and a careful analysis allows for a tentative assignment to different vibrational modes. Quantum chemical electronic structure calculations and group theoretical consideration have been employed to facilitate the interpretation of the experimental data.

In a second part of the thesis, surface and bulk modified diamondoid structures of different sizes with either a thiol functional group or an oxygen inclusion are investigated to determine the influence of targeted chemical modification on the electronic structure. The two different modifications are found to lead to fundamentally different effects. The thiol functional group induces an impurity state that dominates the optical properties and leads to a loss of the size dependence of the optical gap for structures up to 30 carbon atoms. Oxygen inclusion strongly influences the optical response but a size-dependence of the optical gap persists. Both modifications are found to quench the UV photoluminescence of pristine diamondoids.

The present data, taken on atomically defined diamond clusters in the gas phase, reveal for the first time the exact interdependence of the optical response of diamondoids with each of several different structural parameters, such as size, shape and surface functionalization.

Contents

1	**Introduction**	**1**
2	**Diamondoids**	**3**
	2.1 History	5
	2.2 Structure & Nomenclature	9
	2.3 Present Understanding of the Electronic Structure	10
	2.4 Diamondoids From Different Perspectives	14
3	**Theoretical Background and Computational Methods**	**21**
	3.1 The Electronic Structure of Matter	21
	3.2 Molecular Orbital Theory	27
	3.3 Group Theoretical Aspects	34
4	**Experimental Methods**	**43**
	4.1 Diamondoid Samples	44
	4.2 Multi-Purpose Gas Cell	45
	4.3 Synchrotron Radiation	51
	4.4 Optical Absorption Spectroscopy	52
	4.5 Photoluminescence Spectroscopy	56
	4.6 Ultraviolet Photoelectron Spectroscopy	62

5	**Results & Discussion - Part I: Pristine Diamondoids**		**67**
	5.1 Optical Absorption		68
		5.1.1 Nanodiamonds in 1D, 2D, and 3D	68
		5.1.2 *Shape* Not *Size* Defines The Smallest Diamond	77
		5.1.3 Evolution of the Optical Gap	79
	5.2 Photoluminescence		100
6	**Results & Discussion - Part II: Modified Diamondoids**		**123**
	6.1 Adamantane-1-Thiol - The Whole Picture		124
	6.2 Larger Diamondoid Thiols		135
	6.3 Bulk-substituted Diamondoids		148
7	**Summary & Outlook**		**157**

Appendix   161

A Character Tables and Selection Rules   163

B List of Publications   169

List of Tables   173

Bibliography   173

Chapter 1

Introduction

Diamondoids constitute a series of perfectly defined, hydrogen-passivated nanodiamonds. These diamond nanocrystals have become available in various sizes and shapes through their recent isolation from petroleum [1]. Diamondoids always contain a well-defined number of diamond cage units and can therefore be regarded as a series of diamond clusters which, in the macroscopic limit, converges against bulk diamond.

Diamondoids are interesting from various perspectives: With respect to diamond they represent miniaturization in the ultimate, molecular size limit; diamondoids possess technologically interesting properties such as negative electron affinity [2] and are, among other things, used as seeds in CVD diamond growth [3]. The chemical functionalization of diamondoids [4] has paved the way for new diamondoid devices [2] and opens new possibilities for tailoring diamondoid properties. Further, the precise knowledge of each diamondoid's structure affords unprecedented opportunities, such as investigating size and shape effects in nanocrystals or the influence of single impurity atoms.

Typical investigations on neutral clusters and nanocrystals suffer from poorly defined experimental parameters. Especially when studying the optical properties, which prohibits the use of charged particles, a size distribution of the investigated samples is inevitable. Further experimental shortcomings usually include an unknown surface reconstruction, undefined particle shape and interactions of the sample with its environment. The present gas phase investigations on diamondoids allow one to get rid of these experimental nuisances and produce data of atomically defined, interactionless, neutral particles. These are the same boundary conditions of typical theoretical electronic structure investigations. The results of this study therefore provide unrestricted and for nanocrystals unprecedented comparability of experimental and theoretical data.

The goal of the present study is to determine *with atomic precision* the influence of different parameters which define the electronic and optical properties of diamondoids. The investigated parameters are particle size, particle shape, and chemical modification. For this purpose a gas cell setup is developed which facilitates synchrotron-based absorption and photoluminescence measurements on diamondoids in the gas phase. Pristine diamondoids of different sizes and shapes are investigated to determine the influence of size and shape on the optical properties. Diamondoids with surface functional groups and bulk-substituted diamondoids with incorporated impurity atoms are studied to learn about the impact of different kinds of chemical modifications on the electronic and optical properties of diamondoids. Quantum chemical calculations and group theoretical methods are used to help interpret the results.

In chapter 2 diamondoids are introduced in detail and the previous studies on their electronic structure are briefly reviewed. The theoretical and experimental methods which are used in this work are presented in chapters 3 and 4, respectively. Chapter 5 contains the results on the optical properties of pristine diamondoids and in chapter 6 the results for modified diamondoids are discussed. The final chapter summarizes the most important findings of this thesis and gives an outlook on possible interesting developments in the field.

Chapter 2

Diamondoids

Diamondoids are perfectly hydrogen-terminated carbon clusters which can be superimposed on the bulk diamond lattice. They can be thought of as diamond

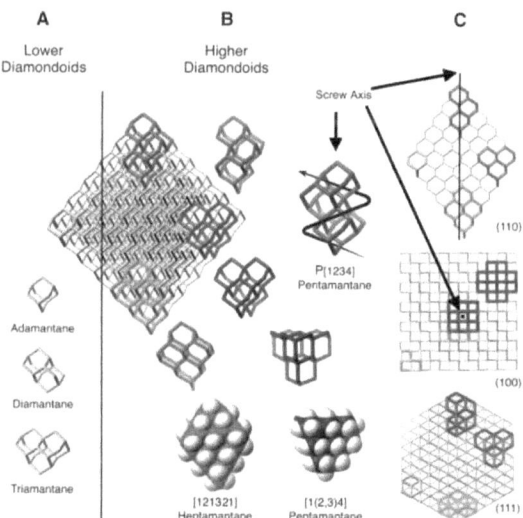

Figure 2.1: Diamondoids can be thought of as (sub-)nanometer diamond fragments. They consist of a small whole number of diamond crystal cages with dangling bonds passivated by hydrogen. Their carbon framework can be superimposed on the bulk diamond lattice as shown. Lower diamondoids have only one, higher diamondoids have multiple isomers. [1]

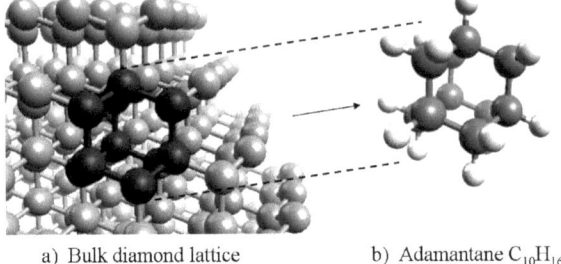

a) Bulk diamond lattice   b) Adamantane $C_{10}H_{16}$

Figure 2.2: Adamantane ($C_{10}H_{16}$) is the smallest closed-cage diamond nanostructure. In a top-down approach it can be constructed by excising a single crystal cell from the bulk diamond lattice (a) and then passivating dangling bonds with hydrogen (b).

fragments which consist of a whole number of diamond crystal cages which are excised directly from the bulk diamond lattice. This is indicated for some diamondoid structures in Fig. 2.1 where only the carbon framework is shown. The dangling bonds are passivated with hydrogen atoms which are omitted in Fig. 2.1 for clarity. While to date experimentally available diamondoids are limited to sizes $\lesssim 1\,\text{nm}$ there is no size limit on principle. Diamondoids therefore form a series of diamond nanocrystals which converges against the bulk diamond structure. As such, they constitute a subcategory of nanodiamond which excels through its perfectly defined surface and its complete sp^3 hybridization.

Adamantane ($C_{10}H_{16}$) is the smallest of the diamondoids consisting of a single diamond cage. In Fig. 2.2 adamantane is constructed in a top-down manner by excising a single diamond cage from the bulk and subsequently passivating the dangling bonds with hydrogen. In diamondoids the bulk diamond structure is conserved and the internuclear distances and angles are in essence identical with the values in bulk diamond. In a bottom-up approach, shown in Fig. 2.3, adamantane can be thought of as a polycyclic hydrocarbon which can be constructed through the attachment of an isobutyl cap to cyclohexane (b) or, alternatively, by fusing four cyclohexane rings to form a cage-structure (c). The example of adamantane demonstrates that diamondoids are in an intermediate regime between hydrocarbon molecules on the one side and diamond (nano-)crystals on the other. This not only refers to their size but also to their composition which shifts with increasing diamondoid size from hydrocarbonaceous more and more towards a 'bulk-like', pure carbon composition due to the decreasing proportion

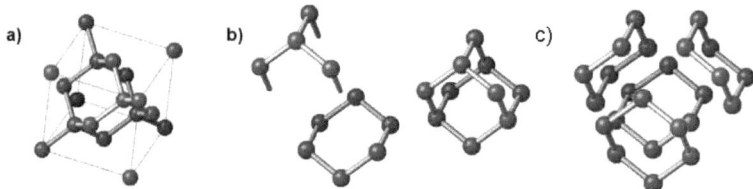

Figure 2.3: The structure of the adamantane carbon framework can be won by (a) removing the atoms at the corners of the diamond unit cell, (b) attaching an isobutyl cap to a cyclohexane ring molecule or (c) fusing four cyclohexane rings. [5]

of surface hydrogen.

2.1 History

Adamantane was first discovered in petroleum in 1933 [6]. In 1941 a successful synthesis was reported [7] but only with the use of a simpler process developed by Paul von Ragué Schleyer in 1957 adamantane became widely available [8]. This novel synthesis route fostered interest in adamantane research and lead to an increasing number of experiments investigating its chemical and physical properties. The surging interest lead to a quest for adamantane homologues, which have later been named diamondoids. For the *XIX International Congress of Pure and Applied Physics*, which took place in 1963 in London, the so far unknown diamantane was chosen as the conference emblem and a challenge for its synthesis was proclaimed. Two years later Cupas and von Schleyer reported the successful synthesis of what they first named *congressane* [9]. The name was one year later officially changed to *diamantane* [10]. Only one year after its first synthesis, in 1966 the synthesis of triamantane followed [11]. In the same year diamantane was first isolated from petroleum [12].
The rapid progress, however, came to a sudden halt when it became clear that the synthesis of higher diamondoids with their more complex structure posed seemingly unsurmountable problems. As of to date, of all higher diamondoids only one of the three tetramantane isomers ([121]tetramantane) was successfully synthesized [13] and as a consequence interest in diamondoids waned in the last part of the 20*th* century.
In the 1990s frail interest reawakened from the field of geophysics when several

CHAPTER 2. DIAMONDOIDS

Figure 2.4: Different diamondoids recrystallized in macroscopic amounts as van-der-Waals crystals. [20]

groups reported the discovery of diamondoids ranging from triamantane [14] to hexamantane [15] in petroleum fields. Diamondoids seem to form naturally in petroleum reservoirs as they have been detected in various places around the world, ranging from Japan [16] and China [17] via Australia [18] to the US mainland [14] and the Gulf of Mexico [15]. They have subsequently been used as an indicator for the maturity, i.e., the degree of biodegradation of a petroleum source [17, 19]. Despite the interest of the oil industry the availability of diamondoids stayed for the time being limited to only the smallest structures.

In 2003 Jeremy Dahl and co-workers reported the isolation of several higher diamondoid structures from petroleum with sizes up to 11 crystal cages [1]. For the isolation Dahl et al. made use of the extreme thermal stability of diamondoids by first thermally decomposing non-diamondoid constituents. Afterwards they applied high-performance liquid chromatography which uses differences in the average geometric cross section to size- and shape-select different diamondoids structures. Polymantanes up to heptamantane have been isolated in macroscopic

2.2. STRUCTURE & NOMENCLATURE

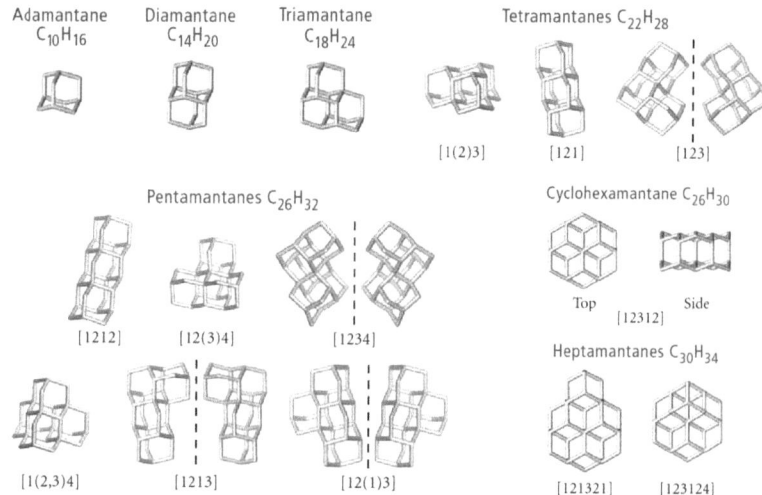

Figure 2.5: Diamondoids of different sizes and shapes [1]. Diamondoids are grouped according to their number of crystal cages. The numbers in square brackets indicate the spatial arrangement of the cages according to the Balaban-Schleyer nomenclature. Only the carbon framework is shown and the hydrogen surface termination is omitted for clarity.

amounts (>1 mg). As such, diamondoids provide a series of nanodiamond structures for experimental investigations which excels through perfectly defined size, shape and surface. At room temperature diamondoids condense in the form of molecular van-der-Waals crystals, shown for some examples in Fig. 2.4. Due to their large band gap diamondoid crystals are colorless and may appear white as a result of the scattering of light. While these molecular crystals have fairly high melting points around 200°C [21] they already sublimate at room temperature [22]. The partial pressure at a given temperature depends on both molecular weight and shape of the diamondoid. The latter is dictating the stacking order within the molecular crystal and therefore the strength of the attractive forces between the single diamondoids which make up the macroscopic molecular crystal. The structures of the single diamondoids are shown in Fig. 2.5.

	B.-S.	Diamondoid	Chemical formula	Molecular point group	Mass [amu]	Bulk atoms C	Surface atoms CH	Surface atoms CH_2
*		adamantane	$C_{10}H_{16}$	T_d	136,23	–	4	6
*	[1]	diamantane	$C_{14}H_{20}$	D_{3d}	188,31	–	8	6
*	[12]	triamantane	$C_{18}H_{24}$	C_{2v}	240,39	1	10	7
*	[121]	tetramantane	$C_{22}H_{28}$	C_{2h}	292,45	2	12	8
*	[123]			C_2		2	12	8
*	[1(2)3]			C_{3v}		3	10	9
	[1231]	pentamantane	$C_{25}H_{30}$	C_s	330,49	2	16	7
*	[1212]		$C_{26}H_{32}$	C_{2v}		3	14	9
*	[1213]			C_1		3	14	9
	[1234]			C_2	344,54	3	14	9
*	[12(1)3]			C_1		4	12	10
	[12(3)4]			C_s		4	12	10
*	[1(2,3)4]			T_d		6	8	12
*	[12312]	hexamantane	$C_{26}H_{30}$	D_{3d}	342,56	2	18	6

Table 2.1: Some properties of selected diamondoids. Structures which are investigated in this work are marked with a '*'. The column labeled 'B.-S.' indicates the structure according to the Balaban-Schleyer notation. 'Bulk atoms' are quarternary carbon atoms, i.e., C atoms only bound to other carbons. 'Surface atoms' are either bound to one or two hydrogen atoms and labeled as CH or CH_2, respectively.

2.2 Structure & Nomenclature

Diamondoid structures are classified by the number of crystal cages. Their name is derived from that of adamantane: It is composed of a Greek numeral prefix indicating the number of cages, or adamantane subunits, followed by the suffix -*mantane*. Diamondoids are therefore sometimes also referred to as *polymantanes*. Diamantane is thus a diamondoid comprising two crystal cages, triamantane comprises three and so on. Diamondoids up to triamantane, for which no structural isomers exist, are also called *lower diamondoids*. For *higher diamondoids* starting with tetramantane, or four diamond cages, different structural isomers for same size diamondoids exist. A reference to the number of cages of a higher diamondoid therefore no longer unambiguously defines its structure. In Fig. 2.5 all diamondoid structures up to tetramantane and a selection of possible pentamantane, hexamantane and heptamantane structures are shown. Combinatory possibilities explode with an increasing number of cages. For tetramantane four structural isomers exist, two of which are enantiomers, for pentamantane there are ten different isomers and for octamantane there are already more than onehundred.

To distinguish these isomers Balaban and von Schleyer introduced a nomenclature that includes the spatial arrangement of the cages: The name of the polymantane is preceded by a numeral prefix (typically in square brackets) which indicates the orientation of the cages along the four axes of a tetrahedron [23]. The concept is schematically shown in Fig. 2.6: The first two cages are necessarily added along two different directions, which are standardly labeled 1 and 2. Note that for diamantane and triamantane the polymantane name is unambiguous and therefore the full notation of [1] diamantane and [12] triamantane is not used. Three different ways exist to attach a fourth diamond cage to triamantane to yield tetramantane. These are shown in the second row of Fig. 2.6: pictogram (e) shows the addition of a cage parallel to the first extension which thus gives rise to [121] tetramantane. In pictogram (f) the additional cage is attached in a different, third direction; therefore the resulting diamondoid is labeled [123] tetramantane. In pictogram (g) the fourth cage branches off the main line. Such branches are indicated in the Balaban-Schleyer nomenclature using round brackets: The diamondoid is identified as [1(2)3] tetramantane. For larger diamondoids another digit is added for each additional cage. For example, for rod-shaped diamondoids the diamondoid name is preceded by an alternating sequence of ones and twos.

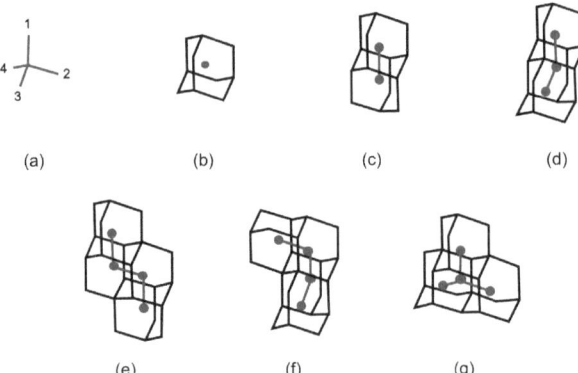

Figure 2.6: Diamondoid structures consist of diamond cages which are fused along the tetragonal axes of the bulk diamond framework. Each additional cage is therefore fused to the preceding structure along one of the four directions indicated in the pictogram (a). Based on the structure of triamantane (d) which defines the first two directions three different tetramantane structures (e)-(g) can be constructed. These are labeled (e) [121] tetramantane, (f) [123] tetramantane and (g) [1(2)3] tetramantane according to the direction and the site of the fourth cage as explained in the text. [23]

2.3 Present Understanding of the Electronic Structure

Until their recent isolation from petroleum diamondoids have not been available and are therefore a largely uninvestigated species. With the isolation of higher diamondoids in 2003 great interest in diamondoids and their electronic structure arose. Experimental work of the recent years includes the investigation of unoccupied [24] and the occupied electronic states [25, 26] as well as a vibrational analysis [27, 28]. It was complemented by several theoretical studies on the electronic structure of diamondoids and small diamond nanocrystals [29, 30, 31]. The experimental results concerning the electronic structure of diamondoids are briefly presented as a background for the discussion in this thesis. Theoretical investigations will be referred throughout this work to where it is deemed helpful.

Adamantane is an exception and has been available to scientists for experiments. Several investigations on the electronic properties [32, 33] exist which in this work serve as a reference for the investigation of larger diamondoids. The investigation

2.3. PRESENT UNDERSTANDING OF THE ELECTRONIC STRUCTURE

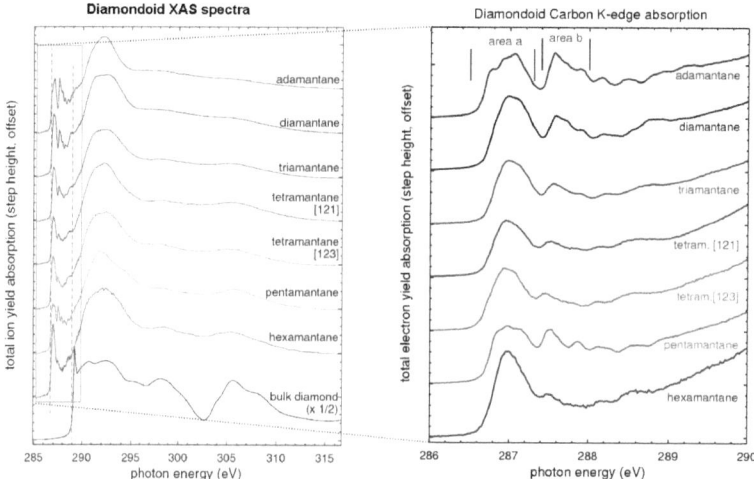

Figure 2.7: X-ray absorption spectra of pristine diamondoids [24]. The right graph shows an enlarged view of the carbon K-edge. The ratio of the intensities of the peaks labeled as *area a* and *area b* is empirically found to match the ratio of CH to CH_2 coordinated surface atoms. The LUMO is therefore ascribed to surface states. The investigated isomers of penta- and hexamantane are [1(2,3)4] pentamantane and [12312] hexamantane (also cyclohexamantane).

of the unoccupied states of diamondoids in 2005 [24] bore a surprise: The energy of the lowest unoccupied molecular orbitals (LUMO) was found to be independent of size which seemed to contradict the well established quantum confinement model. The x-ray absorption spectra of pristine diamondoids are shown in Fig. 2.7. The vertical dotted line demonstrates that the LUMO is fixed in energy, a property that is unique to diamondoids. This behavior is in stark contrast to the behavior of similar semiconductor nanostructures, such as Si- and Ge-nanocrystals, where quantum confinement effects lead to an opening of the band gap with decreasing size due to shifts in both valence and conduction states [34, 35]. The right panel in Fig. 2.7 shows an enlarged view of the carbon K-edge of diamondoids. Among the different diamondoid structures a drastic change in relative intensities of the two main peaks at the absorption edge labeled *area a* and *area b* is observed. The ratio of the spectral intensities of the two peaks is found to correlate precisely with the ration of CH and CH_2 coordinated surface atoms. From these empirical findings the constant energy of the LUMO is concluded to be due to the surface

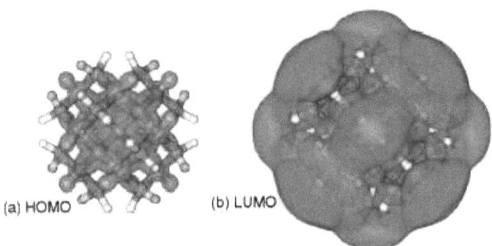

Figure 2.8: Visualization of the electron density distribution for the HOMO and the LUMO of a hydrogen-passivated spherical nanodiamond ($C_{29}H_{36}$) as derived from electronic structure calculations [31].

nature of the lowest unoccupied states. Shortly after, this interpretation was corroborated by electronic structure calculations [31] which found the LUMO to be delocalized around the diamondoid's hydrogen shell and the HOMO to be localized at its core as shown in Fig. 2.8.

Unlike the unoccupied states the highest occupied states of diamondoids exhibit a clear size-dependence. The existence of a shift in the occupied states has first been observed in 2006 using soft x-ray emission [36]. However, this investigation was conducted on condensed diamondoid samples in which the influence of the particle-particle interaction, x-ray beam damage, and other complications remain unclear. A first precise quantification of the size-dependent shift in the occupied states was achieved by the experimental determination of the ionization potentials [25].

The occupied states of diamondoids, valence as well as core levels, were investigated in detail in our group within the diploma thesis of Kathrin Klünder. A summary of the results and a concluding interpretation will soon be published [37]. The photoelectron spectra of a series of pristine diamondoids are shown in Fig. 2.9. The valence band spectra provide an insightful perspective on the size- and shape-dependence of the occupied states of diamondoids. They not only allow for the determination of the valence band edge, respectively the ionization potential, which is demonstrated in the inset in Fig. 2.9. They also reveal similarities and differences in the electronic and vibronic structure of different diamondoids[1]. The different experimental results for the ionization potential, respectively the

---

[1] A detailed discussion of the photoelectron spectra can be found in Ref. [26] and [37].

2.3. PRESENT UNDERSTANDING OF THE ELECTRONIC STRUCTURE

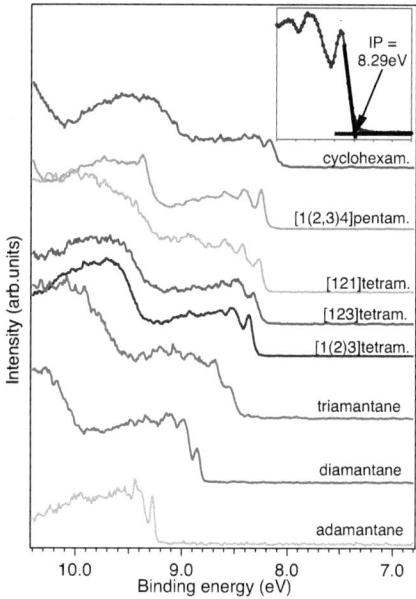

Figure 2.9: Valence band photoelectron spectra for a series of diamondoids. The inset shows the determination of the ionization potential (IP) by the interpolation method using the example of [1(2)3] tetramantane. (To appear in Ref. [37])

valence band edge, are listed in Tab. 2.2 together with some theoretically derived values.

Additionally to the highest occupied states, the C1s core-level binding energies have been measured using x-ray photoelectron spectroscopy. The core-level spectra are shown in the left panel of Fig. 2.10. Here as well, a noticeable, yet much smaller shift towards higher binding energies with decreasing size is observed. A fitting model for the core levels has also been developed. It identifies the contributions of the different chemical environments which allows to separate the effects of screening and of the chemical shift on the measured core-level energies of larger diamondoids. An example for such a core level fit is shown in Fig. 2.10 to the right of the core-level spectra: For [1(2,3)4] pentamantane different chemical environments (C, CH, CH_2) are fitted in the graph corresponding to the ball-and-stick model above. The complex fitting model also accounts for the contributions

Diamondoid	PES (eV)	PIMS (eV)	CL (eV)
adamantane	9,23±0,05	9,23±0,06	289.99
diamantane	8,80±0,05	8,79±0,02	289.91
triamantane	8,44±0,05	8,57±0,01	289.83
[121] tetramantane	8,20±0,05	-	289.71
[123] tetramantane	8,22±0,05	8,11±0,08	289.71
[1(2)3] tetramantane	8,29±0,05	-	289.73
[1(2,3)4] pentamantane	8,18±0,05	8,07±0,03	289.73
[1234] hexamantane	8,00±0,09	-	289.76

Table 2.2: The first ionization potentials as measured by photoelectron spectroscopy (PES) [37] and photoion mass spectrometry (PIMS) [25] and the core ionization potential of the CH_2 chemical environment determined from x-ray photoelectron spectroscopy (CL) [26].

of delocalized vibrational excitations [26, 37]. In the end it allows to deconvolute the different contributions and to determine the core level shifts separately for carbon atoms of different chemical environment. Representatively for the core levels the CH_2 core level binding energies as determined in Ref. [26] are included in table 2.2.

2.4 Different Perspectives: Diamondoids as...

Diamondoids are unique in many ways as they combine the outstanding materials properties of diamond with the tunable properties of nanomaterials. This makes diamondoids a very versatile material system that can be viewed from many different perspectives.

...Novel Form of Nanocarbon

Diamondoids can be regarded as a novel form of nanocarbon which complements the family of nanocarbon materials by adding a well-defined sp^3 nanostructure as shown in the scheme in Fig. 2.11.

2.4. DIAMONDOIDS FROM DIFFERENT PERSPECTIVES

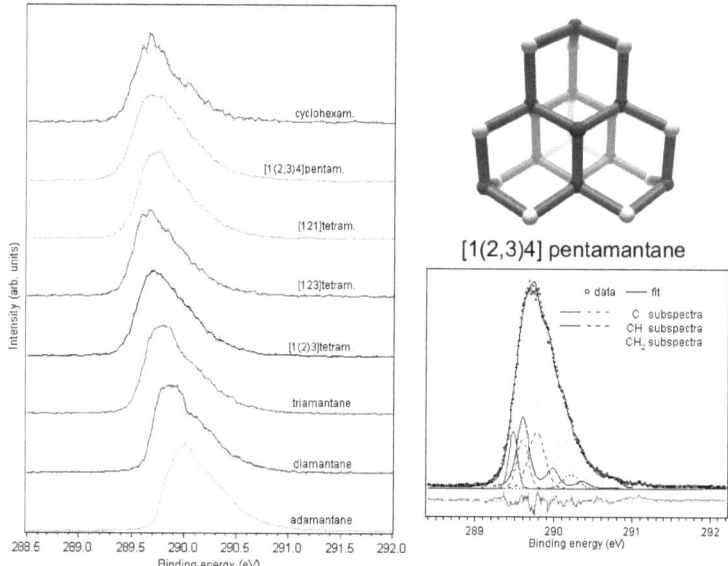

Figure 2.10: Left: The C1s core-level spectra of diamondoids [37]. Right: An example of the core-level fits developed in Ref. [26] (bottom) which deconvolutes the core-level spectra in their single contributions from different chemical environments.

Nanocarbon materials, such as the buckminster-fullerenes, carbon nanotubes (CNTs) or graphene, have caused much excitement within recent years. Most notably, the outstanding importance of these carbon nanostructures for fundamental science and technological research has been recognized by the Nobel Prize committee in Stockholm with a Nobel Prize in chemistry in 1996 for the discovery of the fullerenes and very recently - just a week before the completion of this thesis - the committee announced to award this year's Nobel Prize in physics to Andre Geim and Konstantin Novoselov "*for groundbreaking experiments regarding the two-dimensional material graphene*". In the press release The Royal Swedish Academy of Sciences emphasizes the great relevance of the discovery of the novel nanocarbon material for the scientific and technological development[2] and concludes: "*Carbon, the basis of all known life on earth, has surprised us

---

[2]From the press release: "*Graphene makes experiments possible that give new twists to the phenomena in quantum physics. Also a vast variety of practical applications now appear possible including the creation of new materials and the manufacture of innovative electronics.*" [38]

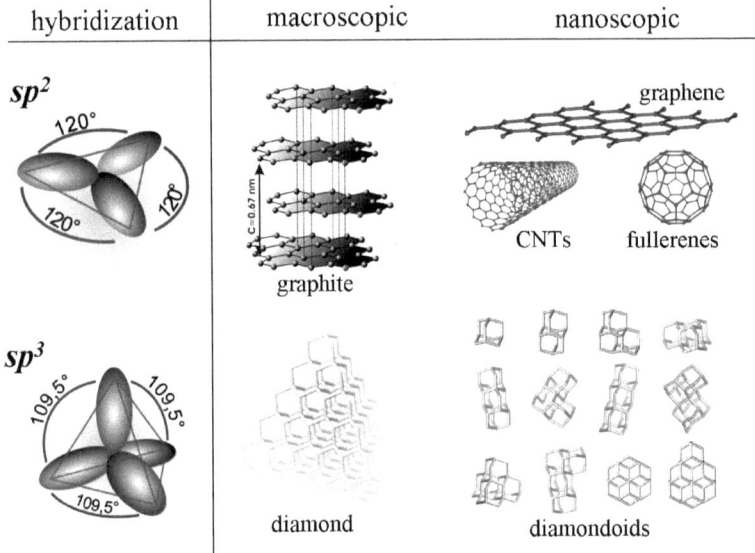

Figure 2.11: Purely sp^2- and sp^3-hybridized forms of carbon. Diamondoids fill in as fully sp^3-hybridized nanocarbon structures. Only the carbon framework is shown and the hydrogen passivation is omitted for clarity.

once again." [38]

Nanocarbon materials excel through a variety of outstanding material properties. Forms of nanocarbon that are based on the sp^2-hybridized structure of graphene exhibit, for example a very high electrical and thermal conductivity. Diamondoids now expand this family by a fully sp^3-hybridized nanostructure which grants access to some of the technologically interesting properties of diamond for the design of combined nanocarbon materials.

...Perfect Semiconductor Clusters

Diamondoids also constitute a series of ideal semiconductor clusters. The notion of "ideal" here, before all, refers to the fact that each diamondoid structure is exactly known in size, shape, and structure. Diamondoids are also referred to as *diamond clusters* due to their structure and the fact that the relevant

2.4. Diamondoids From Different Perspectives

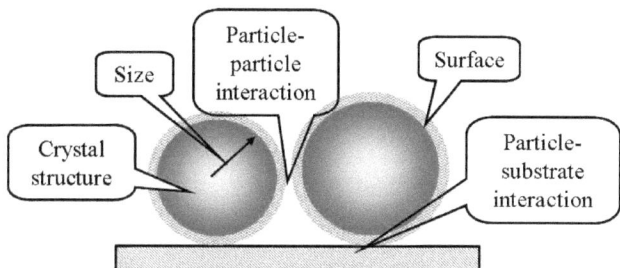

Figure 2.12: Open parameters in typical cluster experiments.

repetitive subunit of diamondoids is a single diamond cage rather than a carbon atom.[3] If continued, the diamondoid series directly converges against macroscopic diamond. The complete hydrogen-termination of diamondoids guarantees a full sp^3-hybridization of the carbon cluster. This leaves no room for uncontrollable surface reconstruction which complicates the production of structurally well-defined clusters. Typically, experiments on clusters have to cope with one or more of the hindrances indicated in Fig. 2.12: the internal structure of the crystal is oftentimes unknown; the samples exhibit a size distribution, especially when investigating neutral particles; the nature of the surface is usually not precisely defined; particle-particle and particle-substrate interactions play an unknown role in experiments on condensed or deposited samples.

All of these experimental nuisances can be circumvented in appropriate experiments on diamondoids. Diamondoids provide a sample system of well-defined, size, shape, and internal structure. The composition and the surface configuration are known with atomic precision. Gas phase investigations on diamondoids further eliminate all particle-particle and particle-substrate interactions. Further, their minuscule size of only a few dozen heavy atoms (<H) makes diamondoids computationally accessible to virtually all *state-of-the-art* theoretical approaches. Interaction-free experiments on these structurally impeccable diamond clusters are therefore suitable to generate experimental data that are directly comparable all current computational electronic structure investigations and to provide a

---

[3]The author remembers an interesting debate during the DyNano2010 conference which evolved among half a dozen cluster scientists over dinner on the question *What makes a particle a cluster?* It was at last (during dessert) agreed on that this question is likely to remain debatable as any overly sharp definition of the term *cluster* will exclude an accepted cluster system.

CHAPTER 2. DIAMONDOIDS

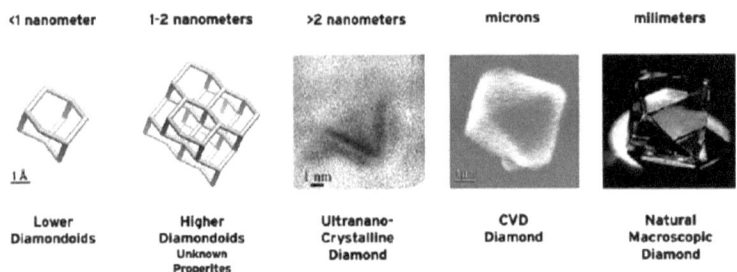

Figure 2.13: Diamondoids represent the smallest possible nanodiamond structures and offer a controlled bottom-up approach to nanodiamond. [41]

benchmark for theory that is of unprecedented quality.

...(Nano-)Diamond

Another perspective on diamondoids is provided by their kinship to macroscopic diamond. Diamond has always fascinated people as gemstone because of its sparkling appearance which is due to its high refractive index. But it is also a material with outstanding mechanical properties. It is the hardest of all bulk materials and it possesses the highest thermal conductivity. Lately, diamond has gained popularity from a very different field: diamond now attracts considerable technological interest for its electronic properties, especially in combination with certain impurities, so-called nitrogen vacancy (NV) centers. This is, before all, due to the fact that NV centers in diamond are promising candidates for the realization of quantum electronic devices including, e.g., solid state quantum computers [39].

Nanodiamond promises to provide the unique properties of diamond in a nano-scale system. Approaches to the synthesis of nanodiamond materials are manifold but they typically are not able to provide nanodiamond particles in sizes smaller than 2-3 nm and the smallest nanodiamonds to contain an NV center measures 5 nm in diameter [40].

Diamondoids with sizes ~1 nm take the miniaturization of diamond a step further to the sub-nanometer level. They are not only the smallest available nanodiamond structures but also the smallest possible ones. Diamondoids also offer a

2.4. DIAMONDOIDS FROM DIFFERENT PERSPECTIVES

bottom-up approach to nanodiamond as shown in Fig. 2.13 and provide a high degree of structural control in size and shape. Even though the available sizes ≤ 1 nm still limit the range of their applications diamondoids offer various advantages over conventional nanodiamonds. E.g. the precise structural control offered by diamondoids may allow to further reduce the size of nanodiamond which contain NV centers. They exhibit perfect bulk diamond structure, internally and on the surface, which makes them ideal nanodiamonds that do not lose diamond properties due to unwanted impurity effects. For this reason the smallest possible diamond will inevitably be a diamondoid structure. Also diamondoids combine the structural properties of bulk diamond with the tunable properties of nanomaterials as they have surpassed the diameter of the exciton Bohr radius in bulk diamond of 1.6 nm [31].

...Novel Materials for Nanotechnology

Diamondoids possess very interesting inherent properties due to their diamond structure, such as bio-compatibility, high thermal stability, and a large band gap. Since the successful isolation of diamondoids from natural oil resources [1] several experimental [24, 36, 42] and theoretical studies [31, 43] have revealed numerous unique and partly unanticipated properties. Some of them have been discovered in the course of this work. Diamondoids exhibit quantum confinement effects that allow their electronic tuning and, in addition to size effects, the shapes of diamondoids strongly influence their optical absorption[4]. Their particular electronic structure leads to highly desirable materials' properties such as negative electron affinity [2, 31] and UV luminescence[4]. The list of interesting properties, which is most likely to be expanded by further investigations, reveals the great prospects that diamondoids bear for use in nanotechnology and various other fields [44].

The growing interest in diamondoids as building blocks for nanotechnology has gained momentum within the last few years due to the successful preparation of surface modified diamondoids [4]. Surface functionalization is a viable method to harness diamondoid properties for technological use. It produces derivatives with potential for applications in the materials sciences, drug development and molecular electronics [45, 46] and also allows for attaching nanoparticles to various other entities such as biomolecules, organic semiconductors, and metal surfaces [47, 48]. Bonding to the latter is typically achieved via thiol functional groups, which form

---

[4]This is a finding of the present work.

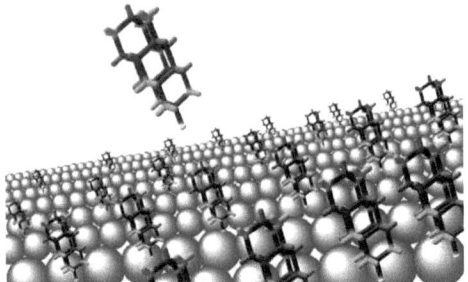

Figure 2.14: Schematic illustration of a self-assembled monolayer of [121]tetramantane-6-thiol on a noble metal surface. [51]

self-assembled monolayers (SAMs) on, e.g., Au- or Ag-surfaces [49, 50]. A variety of diamondoids with different functional groups are now available [41, 52, 53, 54] and further ways to modify diamondoid properties have recently been realized [55] or proposed [56, 57]. Most notably, thiolated higher diamondoids have been prepared [58]. The availability of higher diamondoid thiols now allows for their simple inclusion into solid state devices via SAMs [59, 60], visualized in Fig. 2.14. Recent experimental results [2, 61] provide evidence of the negative electron of diamondoid thiol SAMs affinity and display their great technological potential as low-field electron emitters.

Chapter 3

Theoretical Background and Computational Methods

In this chapter a brief introduction is given into the theoretical understanding of the electronic structure of matter (section 3.1). Also the theoretical tools which are used in this work for the interpretation of the experimental results are presented. In section 3.2 the basics of quantum chemical electronic structure computations are introduced and the underlying theoretical concepts are briefly explained. In section 3.3 a rough-guide to group theory is given which will allow the application of group theoretical considerations to the optical spectra of diamondoids.

3.1 The Electronic Structure of Matter

Diamondoids can be considered a special class of diamond nanocrystals but sometimes also are referred to as *molecular diamonds*. From the viewpoint of cluster physics they are simply hydrogen-passivated carbon or - due to their structure - diamond clusters. These different denominations on diamondoids imply different perspectives and underlying models of description: The nanocrystal is derived top-down from the bulk while the molecule is typically modeled atom by atom. Cluster physics, on the other hand, is trying to bridge the intermediate size regime between molecular and solid state physics and is focussing on the size dependent development of physical properties from a single atom or molecule to the solid. As diamondoids can be justifiably be placed in every one of these fields the basic

CHAPTER 3. THEORETICAL BACKGROUND AND COMPUTATIONAL METHODS

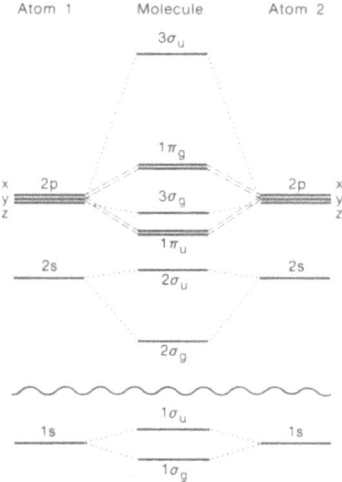

Figure 3.1: Principle of the formation of molecular orbitals for a homonuclear diatomic molecule. The combination of the atomic orbitals leads to a corresponding number of molecular orbitals with bonding and anti-bonding character. Bonding orbitals lie below the original atomic levels, anti-bonding orbitals above them. [62]

ideas of the understanding of the electronic structure of molecules, crystals, and clusters are briefly reviewed.

Molecules

A simple and intuitive model for the description of molecules is derived from the construction of molecular orbitals through the linear combination of atomic orbitals (LCAO). In the LCAO approach, following the variational principle, the coefficients of the linear combination are optimized to yield the minimal total energy of the resulting molecule. For a diatomic molecule, shown in Fig. 3.1, the additive and subtractive superpositions of the atomic wave functions give rise to new, molecular orbitals which differ in their energies. The additive superposition leads to an orbital with increased electron density between the nuclei which thus has a binding effect while the subtractive superposition results in the opposite effect. In the scheme in Fig. 3.1 molecular orbitals with a binding effect lie below the atomic levels while anti-binding orbitals lie above them. They are oftentimes

3.1. THE ELECTRONIC STRUCTURE OF MATTER

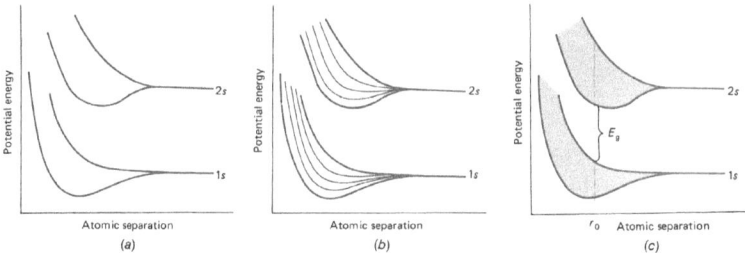

Figure 3.2: The development of a band structure according to the LCAO model. (a) The energy levels for a diatomic molecule. (b) The energy levels for a cluster consisting of five atoms. (c) The quasi-continuous electronic bands of a crystal. The gap energy E_g is given by the vertical distance of two bands at the equilibrium nuclear distance r_0 which is indicated by dashed line. [63]

labeled using a "*", e.g., as σ^* orbital, to distinguish them from binding orbitals. Of particular relevance are the highest occupied molecular orbital (HOMO) and the lowest unoccupied molecular orbital (LUMO). The LCAO approach is also used in quantum chemistry to compute the electronic structure of larger systems. An account of the relevant principles for these calculations is given in section 3.2. In Fig. 3.2 (a) the development of molecular orbitals from the $1s$ and $2s$ atomic levels of a homonuclear diatomic molecule is shown as function of the interatomic distance. The addition of more atoms results in a splitting into multiple levels (panel (b)) which will eventually lead to a quasi continuous band of energy levels in the solid state (panel (c)) [63].

Crystals

In the treatment of the electronic structure of macroscopic crystals the electrons are typically described as wave functions in the periodic potential of the crystal lattice. A particle in a periodic potential can be described as a Bloch wave. The energy of an electron plotted against the wave vector in Fig. 3.3 is known as the electronic band structure. Also the extension of this potential, i.e., the size of the crystal, is assumed to be infinite. Therefore the problem possesses translational symmetry in addition to different kinds of rotational and reflection symmetries which depend on the crystal lattice structure. Due to the translational symmetry it is possible to condense the band structure of a crystal to the first Brillouin zone which results in a reduced zone scheme as shown in Fig. 3.3 (b). The contributions from different Brillouin zones which are stacked in this format give rise to different electronic bands.

The relative position of the energy bands to each other and the degree to which they are occupied by electrons is decisive for the electronic behavior of a material. Overlapping bands or bands which are not fully occupied are typical for metallic behavior because electrons can move about freely, e.g., as a response to an electric field. If the highest populated band is fully occupied and does not overlap with the next higher band their distance is termed the *band gap* of the material. The magnitude of the band gap determines whether a material is considered a semiconductor or an insulator. Typically, materials with small band gaps in the infrared or visible regime of up to a few eV are termed semiconductors. Materials with a larger band gap are considered insulators. The underlying physics, however, is the same for insulators and semiconductors and the delineation between the two is fluent and to a certain degree arbitrary. Diamond for instance is sometimes referred to as an insulator and sometimes as a wide band gap semiconductor. The present investigation of electronic transitions in diamondoids focuses on effects which are typically discussed for semiconductors.

The electronic properties and even more so the optical properties of semiconductors are defined by the band gap between the highest occupied band (valence band) and the lowest unoccupied band (conduction band). The size of the band gap, i.e., the (vertical) distance of the valence band edge from the conduction band

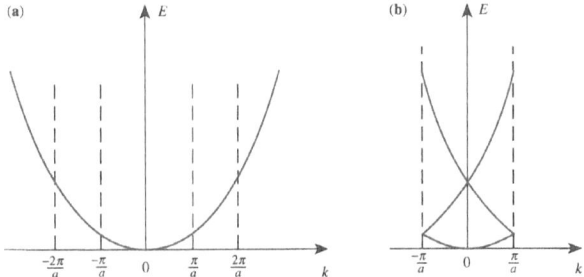

Figure 3.3: Simple representation of the electronic band structure of a crystal: (a) extended zone scheme; (b) reduced zone scheme. [64]

3.1. THE ELECTRONIC STRUCTURE OF MATTER

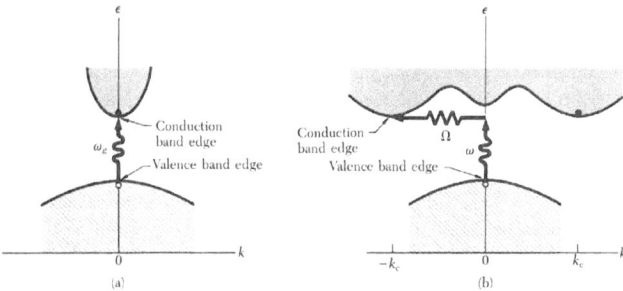

Figure 3.4: Band schemes of (a) a direct band gap semiconductor and of (b) an indirect band gap semiconductor. [65]

edge determines the minimum energy that is required to create an electronically excited state. For optical transitions the relative positions of the valence band edge and the conduction band edge (in the k-space) are also of great importance. Photons carry only negligible momentum compared to electrons and therefore optical transitions occur only vertically in the k-space. As a consequence a band gap can be either *direct* or *indirect*, i.e., the valence and conduction band edge can both occur at the same momentum (typically the Γ point at the center of the Brillouin zone) or at different points in the k-space, respectively. Both cases are sketched in Fig. 3.4. In an indirect band gap semiconductor the direct band-to-band transition is dipole-forbidden due to translational symmetry, in the same way that some transitions in highly symmetric molecules are forbidden within the molecular point group (compare section 3.3). While it only takes a photon with sufficient energy for a direct transition (a), an indirect transition (b) requires the participation of a phonon to account for the missing momentum.

Clusters / Nanocrystals

The introduced electronic band model makes use of the assumption of an infinitely extended periodic potential. This is a very good assumption macroscopic systems. For nanoscopic systems, however, the behavior of electrons starts to notably differ from bulk behavior when the size reaches the order of the exciton Bohr radius. This effect is known as *quantum confinement*.

CHAPTER 3. THEORETICAL BACKGROUND AND COMPUTATIONAL METHODS

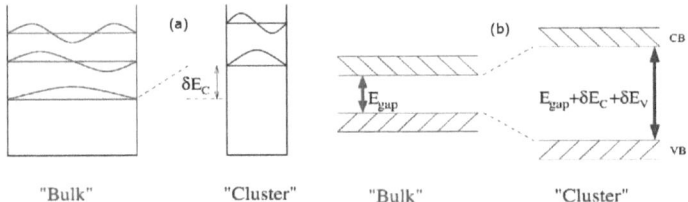

Figure 3.5: Schematic drawing which demonstrates the principle of the quantum confinement effect for a 1-dimensional potential box: If the dimensions are reduced to regime of the de-Broglie wavelength of the electron the energy levels start to shift (a); as a result the band gap of nanocrystals increases with decreasing size (b). [66]

A schematic drawing of the quantum confinement effect for a 1-dimensional box potential is shown in Fig. 3.5. If the size is reduced sufficiently, the continuous energy bands become discrete levels as apparent from Fig. 3.2. This leads to an increased spacing between energy bands and thus to a larger band gap in semiconductor nanocrystals compared to the bulk values. As a technologically interesting effect, the magnitude of the band gap opening can be defined by choosing the particle size. This allows, e.g., to control the light emission properties of nanoparticles as shown in Fig. 3.6 for the example of CdSe quantum dots in solution.

An efficient light emission as observed in CdSe/ZnS quantum dots typically requires materials with a direct band gap. However, the absence of translational symmetry in small nanocrystals can be viewed as a symmetry breaking. As a consequence the electron and hole wave functions in nanocrystals are spread in the k-space, which leads to a breakdown of the k-conservation rules which govern

Figure 3.6: Demonstration of the quantum confinement effect in semiconductor nanocrystals: The light emission of CdSe quantum dots as a function of their size.

26

transitions in ideal macroscopic crystals [67, 68]. Therefore, for sufficiently small nanocrystals it is no longer meaningful to debate whether the gap is direct or indirect.

This understanding of the breakdown of the indirect gap in semiconductor nanocrystals is fairly recent and has come about due to the discovery of photoluminescence from Si-nanostructures. Since the discovery of photoluminescence from Si-nanomaterials in 1990 [69] the prospect of nanosilicon based photonic devices lead to an exploding number of investigations in this field.[1] The concepts which are assumed to underly Si-luminescence, namely a breakdown of the strict momentum conservation in nanocrystals, are of general nature. Still, virtually all of the investigations on the photoluminescence of indirect band gap nanomaterials focus on Si. The same physics, however, should also apply to Germanium and diamond, indirect band gap semiconductors akin to silicon.

3.2 Molecular Orbital Theory

To help the interpretation of the experimental spectra, quantum chemical calculations have been conducted in the course of this work. The quantum chemical program packages GAUSSIAN 03 [70] and TURBOMOLE 6.1 [71] are used in this work to to support experimental investigations. The computed structures are constructed using the corresponding graphical user interfaces GAUSSVIEW and TmoleX, respectively. The results acquired from these electronic structure calculations allow, e.g., to predict values for the energy of electronic levels and to visualize the electron distribution in single molecular orbitals. An example for such a visualization is shown in Fig. 3.7. The reliability of theoretical predictions strongly varies with the employed methods. A brief overview of the theoretical concepts which underlie the computational methods used in this work is given in the following.

Molecular orbital theory predicts the properties of atomic and molecular systems based upon the fundamental laws of quantum mechanics. In the quantum mechanical picture entities posses both particle- and wave-character and are de-

---

[1] The original publication [69] has been cited more than 5000 times as to date.

CHAPTER 3. THEORETICAL BACKGROUND AND COMPUTATIONAL METHODS

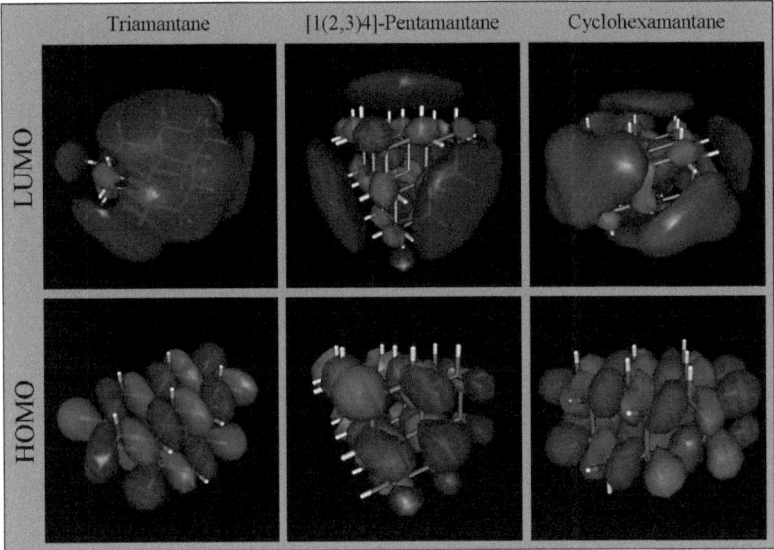

Figure 3.7: Visualization of molecular orbitals. The examples show the computed amplitude of the wavefunction (which corresponds to the electron distributions) in the HOMO and the LUMO for triamantane, [1(2,3)4] pentamantane and [12312] hexamantane.

scribed by the Schrödinger equation:

$$\left\{ \frac{-h^2}{8\pi^2 m} \nabla^2 + V \right\} \Psi(\vec{r}, t) = \frac{ih}{2\pi} \frac{\partial \Psi(\vec{r}, t)}{\partial t} \quad (3.1)$$

Here, Ψ is the wave function, m the mass of the particle, h is Planck's constant and V is the potential field in which the particle is moving. If the potential V is not time-dependent, the Schrödinger equation can be simplified (by separation of variables) to its time-independent form:

$$\mathbf{H} \Psi(\vec{r}) = E \Psi(\vec{r}) \quad (3.2)$$

In this *eigenvalue equation* **H** is the Hamilton operator (compare left side of Eqn. 3.1) and E the *eigenvalue* giving the energy of the particle which is described by the *eigenfunction* $\Psi(\vec{r})$. The pairs of *eigenvalues* and *eigenfunctions* fulfilling this equation for a given molecular system describe the system by giving its stationary states Ψ and the respective energy E.

To describe larger quantum mechanical entities, such as atoms, molecules, or

3.2. MOLECULAR ORBITAL THEORY

nanoparticles, equation 3.1 needs to be applied to a collection of particles. In this case, Ψ will be a function of all particles in the system and of the time t. The potential energy component will then contain a term for the attraction between electrons and nuclei as well as terms for the electron-electron and the nuclear-nuclear repulsion. An exact solution of the Schrödinger equation, however, is not possible for any but the most trivial quantum mechanical systems. Therefore a number of simplifying assumptions are made which help to reach an approximate solution for many-particle systems.

Hartree-Fock

The Hartree-Fock method is an *ab initio* method, i.e., it is based solely on physical constants and does not require any empirical input. The following simplifications are made within the Hartree-Fock method:

- The Born-Oppenheimer approximation is assumed.
- Relativistic effects are completely neglected.
- The mean field approximation is implied, i.e., electron correlation is neglected for electrons of opposite spin.
- The solution is assumed to be a linear combination of basis functions and the finite basis set which is used is assumed to be complete.
- Each energy eigenfunction is assumed to be given by a single Slater-determinant.

In the last point it is assumed that the exact, n-body wave function of a many-body system can be approximated by a single Slater-determinant. The use of a Slater-determinant ensures the antisymmetry which is needed to describe fermionic systems. It is typically composed of the one-particle wave functions for each molecular orbital.

As a next step, to numerically solve the Hartree-Fock-Equations, the molecular orbitals constituting the Slater-determinant are approximated as a linear combination of a pre-defined set of one-electron functions. These so-called *basis functions* could be any set of appropriately defined functions $\chi_1 \ldots \chi_N$, such that the molecular orbital is given by the sum

$$\phi_i = \sum_{\mu=1}^{N} c_{\mu i} \chi_\mu \qquad (3.3)$$

where $c_{\mu i}$ are the molecular orbital expansion coefficients. In this notation ϕ_i refers to an arbitrary molecular orbital and χ_μ to an arbitrary basis function. Typically basis functions are chosen to be centered on the atomic nuclei and so bear some resemblance to atomic orbitals. The molecular orbitals therefore result from a *linear combination of atomic orbitals* (LCAO).

Originally Slater-type orbitals (STOs) were used. For sake of computational simplicity, however, STOs were in turn approximated using linear combination of gaussian-type orbitals (GTOs). GAUSSIAN, TURBOMOLE and other electronic structure programs use gaussian-type atomic functions as basis functions. They are of the form

$$g(\alpha, \vec{r}) = c x^n y^m z^l e^{-\alpha r^2} \tag{3.4}$$

where x, y, and z are the three spatial dimension contained in $\vec{r}$ and α is a constant determining the *radial extent*, i.e., the size of the function. The constant c is chosen to normalize the gaussian function and thus depends on the parameters α, l, m, and n. Standardly, the electronic orbitals of the hydrogen atom are used, e.g., s, p_y, and d_{xy} types:

$$g_s(\alpha, \vec{r}) = \left(\frac{2\alpha}{\pi}\right)^{3/4} e^{-\alpha r^2}$$

$$g_{p_y}(\alpha, \vec{r}) = \left(\frac{128\alpha^5}{\pi^3}\right)^{1/4} y e^{-\alpha r^2} \tag{3.5}$$

$$g_{d_{xy}}(\alpha, \vec{r}) = \left(\frac{2048\alpha^7}{\pi^3}\right)^{1/4} xy e^{-\alpha r^2}$$

Linear combinations of these *primitive gaussians* are then used to form the actual basis functions χ_μ. The final molecular orbital can now be expanded:

$$\phi_i = \sum_\mu c_{\mu i} \chi_\mu = \sum_\mu c_{\mu i} \left(\sum_a d_{\mu a} g_a \right) \tag{3.6}$$

The g_a's denote the gaussian primitives and the $d_{\mu a}$'s are fixed constants within a given basis set.

Basis sets that are used to compute the geometry and the energy levels of molecular systems within the Hartree-Fock framework include minimal basis sets and different kinds of split-valence basis sets. In minimal basis sets each orbital is formed by a single basis function. The most common minimal basis sets are STO-nG where n is an integer giving the number of gaussian primitives to form a basis function. These basis sets are typically used for a first structural optimization.

3.2. MOLECULAR ORBITAL THEORY

They give only rough results which are typically insufficient for research-quality data.

One very commonly used class of *split-valence* basis sets are the basis sets developed in the group of John Pople. The notation of the **Pople basis sets** is typically x-yzG, e.g., the popular 6-31G basis set. In this case the atomic core levels are modeled using a single basis function containing 6 primitive gaussian functions (comp. Eqn. 3.5) and the valence orbitals are composed of two basis functions, one comprising 3 and the other 1 primitive gaussian. The presence of two numbers after the hyphens indicates that this basis set is a *split-valence double-zeta* basis set. The principle of a double-zeta valence basis set is illustrated in Fig. 3.8: A more localized, contracted orbital is combined with a more delocalized, diffuse orbital. This allows to more flexibly model the molecular orbital. 6-311G is a popular example for a split-valence triple-zeta basis set.

Split valence basis sets allow to change the size of the orbital, but not to change its shape. This limitation can be removed by augmenting the basis sets with **polarization functions**. Either d-type polarization functions are added on heavy atoms (>H) only (x-yzG*) or p-functions additionally serve as polarization functions on hydrogen atoms (x-yzG**). The basis sets are also frequently augmented with additional **diffuse functions**. These gaussians which are typically of s or p type have small exponents and, as a consequence, decay slowly with distance from the nucleus. For second row elements they can be regarded as an admixture of 3s- or 3p-orbitals, similar to the augmentation with d-type polarization function. For the Pople basis sets the notation is x-yz+G for the addition of one diffuse s-type and p-type gaussian (with the same exponent) on heavy atoms. The x-yz++G basis sets additionally feature a diffuse s-type gaussian on hydrogens. In this thesis, Pople's basis sets were used almost exclusively.

Another class of basis sets which shall briefly be mentioned for its widespread use is the class of the **correlation-consistent basis sets**. For first and second-row elements, the notation is cc-pVNZ with $N = D, T, Q, \ldots$ for double, triple, quadruple, ... zeta, respectively. Here, cc-p stands for *correlation-consistent polarized* and 'V' indicates that these basis sets are 'valence-only' basis sets. Augmented versions including diffuse functions exist as well and are denoted aug-cc-pVNZ. These basis sets, which are more complex than the Pople's basis sets, have been used to selectively check for consistency with the calculations at some points.

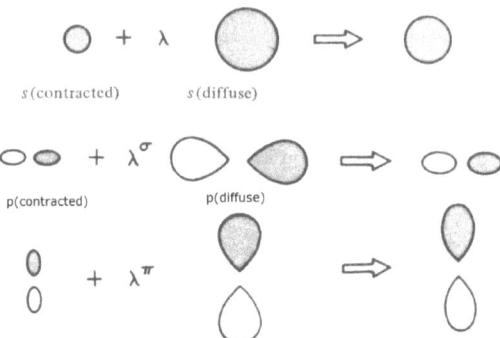

Figure 3.8: Principle of a split-valence basis set: Two (or more) basis functions with different radial extent are combined to yield an orbital that is variable in size [72].

As mentioned above, Hartree-Fock theory provides an inadequate treatment of the electron-electron interaction, especially between electrons of opposite spin. Therefore, electron correlation is oftentimes included giving rise to post-Hartree-Fock methods. Prevalent post-Hartree-Fock methods include, e.g., configuration interaction (CI) and Møller-Plesset perturbation theory (MP2, MP3, MP4,...) and coupled cluster (CC). The more accurate description of the molecular system, however, comes at the cost of high computational demand.

An alternative to post-Hartree-Fock methods which comes at lower computational cost is **density functional theory** (**DFT**). DFT models the electronic structure of a given system as a functional of the electron density; therefore its name. Published in 1964, the Hohenberg-Kohn theorem demonstrated the existence of a functional which determines the ground state energy and electron density exactly. Little does it say, however, about the form of this functional. Following on the work of Kohn and Sham, current DFT methods partition the electronic energy into several terms:

$$E = E^T + E^V + E^J + E^{XC} \qquad (3.7)$$

where E^T is the kinetic energy of the electrons, E^V incorporates the potential energy of the electron-nuclear attraction and the repulsion between the different nuclei, E^J is the electron-electron repulsion term (oftentimes also labeled 'Coulomb self-interaction' of the electron density), and E^{XC}, finally, is the exchange-correlation

3.2. MOLECULAR ORBITAL THEORY

term which includes the remaining part of the electron-electron interactions. The sum of the first three terms corresponds to the classical energy of the charge distribution ρ. The exchange-correlation term accounts for the exchange energy arising from the antisymmetry of the quantum mechanical wavefunctions and dynamic correlations in the motions of the individual electrons. It is therefore usually split up into an exchange and a correlation part:

$$E^{XC}(\rho) = E^X(\rho) + E^C(\rho) \tag{3.8}$$

As indicated in Eqn. 3.8, the *exchange functional* $E^X(\rho)$ as well as the *correlation functional* $E^C(\rho)$ are again functionals of the electron density ρ. The problem in DFT is, that these are not exactly known but for the free electron gas. Approximations have to be made in order to derive physical quantities.

One of the most common approximations is the local-density approximation (LDA), where the functional at a certain point in space only depends on the electron density at this point. The generalized gradient approximation (GGA) also takes into account the gradient of the electron density at that point.

In actual practice, self-consistent Kohn-Sham DFT calculations are performed in an iterative manner that is analogous to Hartree-Fock methods. This allows for the formulation of hybrid functionals which include a mixture of the exact Hartree-Fock exchange along with DFT exchange and correlation:

$$E^{XC}_{hybrid} = c_{HF} E^X_{HF} + c_{DFT} E^{XC}_{DFT} \tag{3.9}$$

where the c's are constant parameters which can be determined empirically. For example, the popular Becke-style three-parameter B3LYP functional, which is also used in this work, is defined as:

$$E^{XC}_{B3LYP} = E^X_{LDA} + c_0(E^X_{HF} - E^X_{LDA}) + c_X \Delta E^X_{B88} + E^C_{VWN3} + c_C(E^C_{LYP} - E^C_{VWN3}) \tag{3.10}$$

Using this approach, the parameter c_0 allows any admixture of Hartree-Fock and LDA, i.e., DFT local exchange to be used. Becke's gradient correction B88 using GGA is included as well as the VWN3 local correlation functional. The latter may be corrected by the LYP correlation correction introduced by Lee, Yang, and Parr [73]. In the B3LYP functional, the open parameters have been determined empirically by Becke by fitting a set of predicted to measured values for ionization potentials, total atomic energies, etc. They are $c_0 = 0.20$, $c_X = 0.72$, and $c_C = 0.81$.

CHAPTER 3. THEORETICAL BACKGROUND AND COMPUTATIONAL METHODS

3.3 Group Theoretical Aspects

Group theory provides a powerful tool to derive physical properties of a system from its mere symmetry and without any further, more specific knowledge of its structure or composition. In this thesis, group theoretical considerations are combined with electronic structure calculations to identify dipole forbidden transitions in the optical spectra of diamondoids.

In the following, a brief overview of the group theoretical fundamentals and their application to optical transitions, i.e., to electric dipole transitions, is given. A comprehensive treatment of the underlying mathematical concepts, however, is beyond the scope of this thesis and some fundamental mathematical concepts, such as groups, are assumed to be known. For a full account the interested reader is referred to the standard literature.[2]

Molecular Point Groups

Group theory allows to infer physical properties from mere knowledge of the symmetry of the investigated system. The symmetry of a system, e.g., of an equilateral triangle shown in Fig. 3.9, is defined by the set of symmetry operations which leave the system unchanged. Examples for such symmetry operations are a rotation by 120° (3-fold symmetry) or the mirroring on an axis. The operators defining these symmetry operations form a group which is called a *point group* if at least one point is fix under all operations. For quantum chemical problems it is useful to limit the reflections to the three-dimensional, so-called *molecular* point groups.

The point group of a system specifies all possible symmetry operations. Point groups are classified according to the symmetry elements they contain. A symmetry element can, e.g., be an n-fold rotational axis (C_n), a mirror plane (σ), or an inversion center (i). The most common molecular point groups and the corresponding symmetry elements are listed in table 3.1.

In the case of a system with the symmetry of the equilateral triangle (contem-

---

[2] An excellent and yet brief review of the necessary concepts can be found in the book *How To Use Groups* by J. W. Leech and D. J. Newman [74]. A more comprehensive account, suitable for scientists who are unacquainted with group theory, can be found in Ref. [75] (German).

3.3. GROUP THEORETICAL ASPECTS

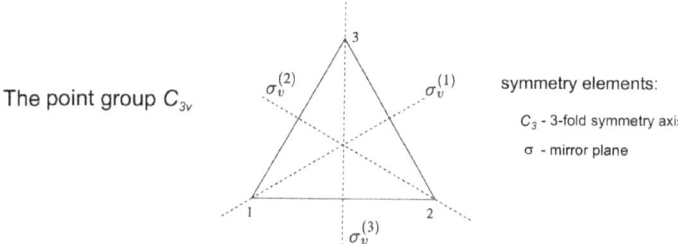

Figure 3.9: The symmetry of the equilateral triangle (in the plane) as an example of the point group C_{3v}. The diamondoid [1(2)3] tetramantane is of the same symmetry.

Point group	Symmetry elements
C_n	n-fold rotation axis
C_{nv}	n-fold axis + n mirror planes containing the axis
C_{nh}	n-fold axis + a perpendicular mirror plane
	(+ an inversion center i if n is even)
D_n	n-fold axis + n 2-fold axes (perpendicular to the principal axis)
D_{nd}	like D_n, additionally n mirror planes containing the principal
	axis and bisecting the angle between two of the 2-fold axes
D_{nh}	like D_n + a mirror plane perpendicular to the principal axis
T_d	all symmetry elements of the regular tetrahedron
O_h	all symmetry elements of the regular octahedron
I_h	all symmetry elements of the regular icosahedron
Note:	$C_s \equiv C_{1v} \equiv C_{1h} \equiv S_1$; $C_i \equiv S_2$

Table 3.1: The most common molecular point groups and their symmetry elements.

CHAPTER 3. THEORETICAL BACKGROUND AND COMPUTATIONAL METHODS

plated only in the plane), such as [1(2)3] tetramantane, the corresponding point group is C_{3v}. The symmetry operations within this group can be labeled as follows: E - the identity; A_1 - a 120° rotation about the central axis; $A_2 = A_1^2$ - a 240° rotation about the central axis; D_1 - a reflection in the plane $\sigma_v^{(1)}$; $D_2 = A_1 D_1 = D_1 A_2$ - a reflection in the plane $\sigma_v^{(2)}$; $D_3 = A_2 D_1 = D_1 A_1$ - a reflection in the plane $\sigma_v^{(3)}$. As can be seen from the equalities between certain operations the subsequent application of two operators results in a different operator of the group. The interrelation of all symmetry operations within a point group is in the literature typically summarized in a *multiplication table*.

Irreducible Representations

For the application of group theory it is necessary to study special mathematical *representations* of the group elements. A mathematical representation formalizes the behavior of the system under the application of a certain element of its symmetry group. Many different possibilities, such as a representation by permutations, exist but matrix representations proof particularly useful. A matrix representation of a group $\mathcal{G}$ assigns a matrix $\Gamma(R)$ to every $R \epsilon \mathcal{G}$ such that

$$\Gamma(A)\,\Gamma(B) = \Gamma(C) \text{ whenever } AB = C.$$

Within the C_{3v} point group the operation A_1 (rotation by 120°) can, e.g., be represented using a 6×6 matrix which produces a rearrangement, or permutation, of the operators:

$$A_1 \begin{bmatrix} E \\ A_1 \\ A_2 \\ D_1 \\ D_2 \\ D_3 \end{bmatrix} = \begin{bmatrix} A_1 E \\ A_1 A_1 \\ A_1 A_2 \\ A_1 D_1 \\ A_1 D_2 \\ A_1 D_3 \end{bmatrix} = \begin{bmatrix} A_1 \\ A_2 \\ E \\ D_2 \\ D_3 \\ D_1 \end{bmatrix} = \begin{bmatrix} 0 & 1 & 0 & 0 & 0 & 0 \\ 0 & 0 & 1 & 0 & 0 & 0 \\ 1 & 0 & 0 & 0 & 0 & 0 \\ 0 & 0 & 0 & 0 & 1 & 0 \\ 0 & 0 & 0 & 0 & 0 & 1 \\ 0 & 0 & 0 & 1 & 0 & 0 \end{bmatrix} \begin{bmatrix} E \\ A_1 \\ A_2 \\ D_1 \\ D_2 \\ D_3 \end{bmatrix}$$

The transpose of this matrix, rather than the matrix itself[3], is a particular representation of A_1. Different representations are possible, e.g., a representation based on the effect of the operations on the vector $\vec{r} = (x, y, z)$.

In fact, there is an infinite number of representations which fulfil the above requirement. But any given group possesses only a limited number of so-called

---
[3] Transposition is necessary to preserve the order of multiplication.

3.3. GROUP THEORETICAL ASPECTS

irreducible representations. From these irreducible representations all other representations can be constructed and, in consequence, any arbitrary representation Γ of a group $\mathcal{G}$ can be decomposed into a direct sum of irreducible representations. Note that the irreducible representations play a key role for the application of group theory to quantum mechanical systems. Most notably, molecular orbitals of a particular system transform under symmetry operations as one of the irreducible representations of the corresponding molecular point group.

The theory of group characters provides a simple technique to decompose an arbitrary representation into its irreducible components. The characters $\chi(R)$ of a representation Γ of a group $\mathcal{G}$ are defined as the trace of their representation matrices $\Gamma(R)$:

$$\chi(R) = Tr\Gamma(R) = \sum_i^d \Gamma_{ii}(R) \qquad (3.11)$$

for all operations R where d is the dimension of the representation. The characters of the irreducible representations of a group are usually summarized in a *character table*. A character table is organized as a matrix and lists the characters of symmetry operations with respect to each of the irreducible representations of a group. The character table for the C_{3v} point group is the following:

C_{3v}	E	$2C_3$	$3\sigma_v$
A_1	1	1	1
A_2	1	1	-1
E	2	-1	0

In a character table all symmetry operations which have the same characters for all irreducible representations are pooled in *classes*. Each class is labeled by a typical element and the classes are listed in the first row of the table. In the above example of the C_{3v} point group the classes are E, C_3, and σ_v containing 1, 2, and 3 elements, respectively. Their characters with respect to each irreducible representation are given below in the corresponding line. Most notably, the dimension of the irreducible representation is given by the character of the identity operation E (which always constitutes a class of its own and is not to be confused with the irreducible representation E).

Character tables for all relevant point groups exist in the literature and are listed in the Appendix. For a better understanding of the implications of characters it is helpful to know some conventions for the notation of irreducible representations:

(i) Representations labeled A and B are 1-dimensional; the label A is used if the

CHAPTER 3. THEORETICAL BACKGROUND AND COMPUTATIONAL METHODS

character corresponding to the principal rotation C_n is 1, if it is -1 the label B is used.

(ii) Representations named E are 2-dimensional.

(iii) T representations are 3-dimensional. There are no irreducible representations of the molecular point groups with dimensions higher than three.

(iv) The subscripts g and u indicate even and odd representations under inversion, respectively.

The dimension of an irreducible representation, which can be read out directly from the character table of the corresponding point group, is of particular interest for quantum mechanics. In the treatment of molecular orbitals it indicates the degeneracy of all orbitals which transform like this irreducible representation.

Further, character tables can be used to decompose an arbitrary representation into a combination of irreducible representations. For this purpose, first, the characters of each class are determined for the representation according to Eqn. 3.11. These characters can be considered a vector. Similarly, for each of the irreducible representations the characters are read out from the character table as a row vector. These vectors for the irreducible representations fulfil the *orthogonality relation*, i.e., they are linearly independent. This allows to decompose the character vector of the arbitrary representation into a linear combination of the irreducible ones. An arbitrary representation R of the C_{3v} point group could have, e.g., the following characters:

	E	$2C_3$	$3\sigma_v$
$\chi(R)$	4	1	-2

Comparison with the character table of the C_{3v} point group shows that $\chi(R) = \chi(E) + 2\chi(A_2)$. The representation R can thus be reduced to a direct sum of the irreducible representations E and A_2 (the latter with multiplicity 2).

In combination with dipole selection rules, which are introduced next, the decomposition of an arbitrary representation will allow to identify forbidden transitions between molecular orbitals of different symmetries.

Selection Rules

As a next step the coupling of two quantum mechanical states $|i\rangle$ and $|j\rangle$ which are eigenvectors of a system to a particular Hamiltonian $\mathcal{H}$ is regarded. The coupling

3.3. GROUP THEORETICAL ASPECTS

of $|i\rangle$ and $|j\rangle$ under an operator $\mathbf{K}$ is given by $\langle i|\mathbf{K}|j\rangle$. From group theoretical considerations and without explicit calculations it can now be concluded that certain of these matrix elements will be zero. Such predictions are known as *selection rules*.

The eigenvectors $|i\rangle$ and $|j\rangle$ of $\mathcal{H}$ are basis functions of two irreducible representations, Γ_α and Γ_β, respectively, of the symmetry group $\mathcal{G}$ of the system. The eigenvectors can be written as $|i\rangle = |\alpha r\rangle$ and $|j\rangle = |\beta s\rangle$ where r and s are particular elements of the bases of the representations Γ_α and Γ_β. The selection rules are then simply derived from the orthogonality relations of the basis functions:

$$\langle \alpha r|\beta s\rangle = \delta_{\alpha\beta}\delta_{rs} \qquad (3.12)$$

The quantum operator $\mathbf{K}$ transforms under all of the symmetry operations of $\mathcal{G}$ as a basis element of some representation Γ_K. This representation will not normally be irreducible. In this case $\mathbf{K}|\beta s\rangle$ is a basis element of $\Gamma_K \otimes \Gamma_\beta$[4] and, applying equation 3.12, one derives the *fundamental selection rule*:

$\langle \alpha r|K|\beta s\rangle$ can only be non-zero if $\Gamma_K \otimes \Gamma_\beta$ contains Γ_α.

Equivalently, it can be demanded that $\Gamma_\alpha \otimes \Gamma_K \otimes \Gamma_\beta$ contain the identity representation. This fundamental selection rule only considers the $\delta_{\alpha\beta}$ part of equation 3.12. It therefore only states whether, in the whole range of possible values for r and s, non-zero elements for $\langle \alpha r|K|\beta s\rangle$ exist. However, all state vectors $|\alpha r\rangle$ ($|\beta s\rangle$) describe states of the same energy α (β). The fundamental selection rule thus determines if *any* coupling between two states with the energies α and β is possible. It does not provide any hint on how strong the coupling might be. This means the derived selection rules allow, in particular, to determine whether optical transition can occur but do not allow to predict their oscillator strength.

The fundamental selection rules derived above are of general nature. To apply them to optical transitions, i.e., to the interaction of matter with an electromagnetic field, the operator describing the coupling has to be defined. The Hamiltonian describing the response of a system to an electromagnetic field $\mathbf{E}$ can be approximated using perturbation theory. In a first order approximation only the largest term is considered which is the electric dipole term.[5] Thus, to derive the

---

[4]$\otimes$ denotes the direct product. The direct product of two groups $\mathcal{G}_1 = \{A\}$ and $\mathcal{G}_2 = \{B\}$ forms a new group $\mathcal{G}_1 \otimes \mathcal{G}_2$ which contains all product operators AB.

[5]The next biggest term is the magnetic dipole term which is on the order of 10^{-5} weaker.

CHAPTER 3. THEORETICAL BACKGROUND AND COMPUTATIONAL METHODS

dipole selection rules the matrix elements of the dipole operator $\mathbf{E} \cdot \mathbf{D}$ have to be considered. These are, basically, the components of the momentum operator $\mathbf{p}$ which transform like the vector $\mathbf{r} = \begin{pmatrix} x \\ y \\ z \end{pmatrix}$, where x, y, and z are the position operators for the three spatial dimensions. Only transitions with a non-vanishing matrix element in $\langle a|\mathbf{r}|b\rangle$ have a finite probability for a dipole transition. According to the fundamental selection rule an interaction requires the irreducible representations of initial and final states to be coupled via the interaction operator. An interaction with the dipole operator can occur in the form of a coupling with any of its three spatial components, i.e., with any of the position operators[6]. To calculate possible dipole interactions the irreducible representation of the initial state R_i is multiplied with the irreducible representations $\Gamma_{x,y,z}$ of the position operators x, y, z:

$$R_i \otimes \Gamma_{x,y,z} = R_F$$

The result is a (in general reducible) matrix R_F. In order for a transition to be dipole-allowed, the irreducible representation R_f of the final state has to be contained in $R_i \otimes \Gamma_{x,y,z}$ where $\Gamma_{x,y,z}$ is generated by the three position operators. The direct product can be decomposed into the irreducible representations R_f of all final states to which the initial state couples. The irreducible representations $\Gamma_{x,y,z}$ of the three position operators for each point group can be found in the character tables (see Appendix).

To apply the fundamental selection rules to a system of a particular point group $\mathcal{G}$ the direct product of those irreducible representations which generate the components of the dipole operator $\mathbf{E} \cdot \mathbf{D}$ with each irreducible representation of the group $\mathcal{G}$ has to be formed. In practice, the characters of the irreducible representations are multiplied and the result is decomposed. This is exemplified in the following using the point group C_{3v}:

The C_{3v} point group has the irreducible representations A_1, A_2, and E. Further, x and y transform as E, and z transforms as A_1. Application of the selection rule identifies which transitions may occur from states of each irreducible representa-

---

[6]Physically this implies non-polarized light with respect to the alignment of the investigated system. Since the experiments in this work are conducted in the gas phase a polarization of the light source does not have an effect on the experimental results.

3.3. GROUP THEORETICAL ASPECTS

tion:

$$\begin{aligned}
\text{transitions from } A_1: A_1 \otimes A_1 &= (1\ 1\ 1) \times (1\ 1\ 1) = (1\ 1\ 1) = A_1 \\
A_1 \otimes E &= (1\ 1\ 1) \times (2\ -1\ 0) = (2\ -1\ 0) = E \\
\text{transitions from } A_2: A_2 \otimes A_1 &= (1\ 1\ -1) \times (1\ 1\ 1) = (1\ 1\ -1) = A_2 \\
A_2 \otimes E &= (1\ 1\ -1) \times (2\ -1\ 0) = (2\ -1\ 0) = E \\
\text{transitions from } E: E \otimes A_1 &= (2\ -1\ 0) \times (1\ 1\ 1) = (2\ -1\ 0) = E \\
E \otimes E &= (2\ -1\ 0) \times (2\ -1\ 0) = (4\ 1\ 0) = A_1 \oplus A_2 \oplus E
\end{aligned}$$

The forbidden transitions are those which do not result from any of the direct products. Thus, from simple group theoretical considerations follows: $A_1 \not\to A_2$ and $A_2 \not\to A_1$. This means that for a physical system with C$_{3v}$ symmetry, like [1(2)3] tetramantane, optical transitions between electronic orbitals which transform as the irreducible representations A_1 and A_2 are dipole-forbidden ($A_1 \not\leftrightarrow A_2$). Dipole forbidden-transitions are summarized in the following table:

	$A_1 \stackrel{\wedge}{=} z$	$E \stackrel{\wedge}{=} x, y$	Forbidden transitions
A_1:	$A_1 \otimes A_1 = A_1$	$A_1 \otimes E = E$	$A_1 \to A_2$
A_2:	$A_2 \otimes A_1 = A_2$	$A_2 \otimes E = E$	$A_2 \to A_1$
E:	$E \otimes A_1 = E$	$E \otimes E = A_1 \oplus A_2 \oplus E$	–

Such tables summarizing the dipole-forbidden transitions have also been worked out for all other relevant point groups. They can be found together with the corresponding character tables in the Appendix.

Chapter 3. Theoretical Background and Computational Methods

Chapter 4

Experimental Methods

In this chapter the experimental methods are described which have been used to investigate the electronic structure of diamondoids. The focus of this thesis lies on the investigation of the optical properties of diamondoids and its derivatives. This thesis provides experimental data that can be directly compared to theoretical studies. Typical quantum chemical computations of the electronic structure assume single, isolated particles. In addition, a particle investigated by theory is, naturally, completely defined in size and shape. The strict monodispersity of the diamondoid samples and the precise knowledge of their size and shape allows to emulate these conditions. For this purpose all the experimental investigations were performed in the gas phase. This avoids interaction of the diamondoids with their surroundings as well as among each other [76]. In combination with the knowledge of the sample structure this allows for a direct comparison of the experimental data to theoretically derived values.

The preparation procedures of the samples are described in section 4.1. The structures of the investigated diamondoid and their general properties were introduced in chapter 2. Optical absorption spectroscopy and photoluminescence spectroscopy provide the main experimental tools of this study. They are described in sections 4.4 and 4.5, respectively. Both these methods use a gas cell setup which is described in section 4.2 and synchrotron radiation described in section 4.3. Further, ultraviolet photoelectron spectroscopy has been employed to yield insight into the highest occupied states of diamondoid thiols (which are discussed in section 6.2). The experimental setup is described in section 4.6.

CHAPTER 4. EXPERIMENTAL METHODS

4.1 Diamondoid Samples

Pristine Diamondoids

Adamantane is commercially available and has been purchased in high purity (99^+%) from Sigma-Aldrich [77]. All other diamondoid samples used in this work were acquired from our collaboration partners from MolecularDiamond Technologies / Stanford University. The diamondoids were isolated and purified from petroleum [1]. Higher diamondoids were obtained by vacuum distillation above $345°$ C and are pyrolyzed at $400°$ to $450°$ C to remove non-diamondoid hydrocarbons. This sophisticated extraction is required because higher diamondoids have to date not been synthesized [44] and no other sources for higher diamondoids are known to date. Also, this procedure, which makes use of the high thermal stability of diamondoids, demonstrates that thermal dissociation of diamondoids does not have to be feared in the comparably low temperature regions ($< 250°$C) used in this work. Size and shape selectivity occurred via high-performance liquid chromatography [1]. All samples exhibit purities exceeding 99% and their structures are confirmed by single crystal x-ray diffraction.[1]

Available sample quantities range from several grams for adamantane and diamantane to only a few mg for the pentamantanes or cyclohexamantane. This scarcity of several higher diamondoid structures is one of the experimental challenges in this work. It requires the different experimental setups to work very efficiently as entire spectra have to be gained from such minimal amounts of the samples.

Modified Diamondoids

Thiol functionalized diamondoid samples used in this work were acquired through a collaboration with the group of Prof Peter R Schreiner from the Justus-Liebig-University in Giessen, Germany. Pristine diamondoids, provided through MolecularDiamond Technologies, serve as starting point and are subsequently chemically modified [58, 41, 53, 4]. The chemical processes exhibit yields which are typically clearly below 100%. Therefore, sample amounts are even more critical for functionalized diamondoids than for pristine ones. The possibility of thermal dissociation at the applied temperatures in this work can also be excluded for

---
[1] A detailed description is given in Ref. [1]

4.2. MULTI-PURPOSE GAS CELL

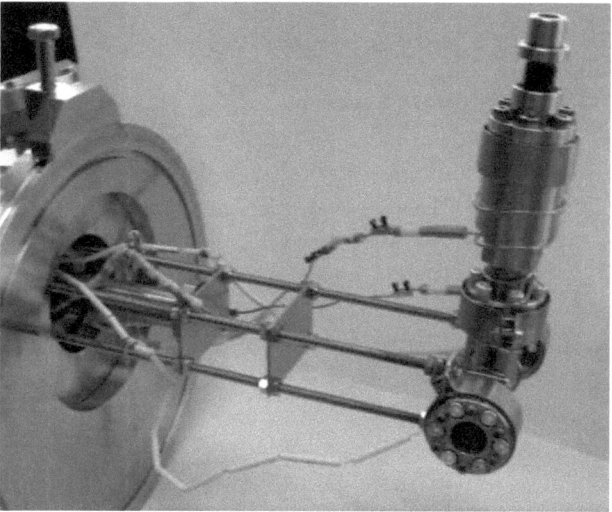

Figure 4.1: Picture of a gas cell setup which was constructed for measuring the absorption properties of diamondoids.

functionalized diamondoids [78].

Further, bulk-substituted diamondoids, i.e. diamondoids for which carbon atoms have been replaced with different elements, are investigated. The model systems investigated in this work are a series of oxadiamondoids and urotropine. Urotropine (or hexamethylenetetramine, $C_6N_4H_{12}$), an adamantane cage for which the tertiary carbon atoms have been replaced by nitrogen, is available commercially and has been bought at Sigma-Aldrich [77]. In oxadiamondoids one binary carbon is replaced by an oxygen atom. The oxa-versions of adamantane through triamantane have been provided through the collaboration with the group of Prof. Schreiner [55].

Just like for pristine diamondoids, sample purities for all of the functionalized species clearly exhibit 99% [77, 78].

4.2 Multi-Purpose Gas Cell

To measure the absorption and photoluminescence of diamondoids in the gas phase a heatable gas cell setup has been developed. The cell design is based upon a prototype that was constructed for previous absorption measurements [79]. A photograph of this primary absorption cell design is shown in Fig. 4.1. Within the last three and a half years it has been continuously advanced to meet the demands of new experimental ideas. Most notably, it was redesigned to allow for the detection of absorption and photoluminescence. The resulting multi-purpose setup not only enables different experiments within the same cell-setup it also allows the parallel measurements of absorption and photoluminescence for an efficient use of the synchrotron beamtime.

In its latest design, which is shown in Fig. 4.2, the gas cell has been adapted to, next to absorption scans, enable spectrally and time-resolved photoluminescence measurements. For this purpose a CF 16 cube is used as main body of the cell. The axis of the cube which is oriented along the beam direction is extended by a 76 mm CF 16 tube for a total cell length of 12.7 cm. This provides an absorption length that is comparable to the original absorption setup [79] and does, therefore, not compromise the suitability of the new multi-purpose setup for the absorption measurements.

Compared to the primary setup the cube allows for the attachment of additional windows. In its current version the cell features four windows, all in the horizontal plane: An entrance and an exit window in beam direction and two windows at $90°$ angles for the detection of photoluminescence. On the bottom face of the cube a blind flange is attached. The use of a cube at the front end of the cell also ensures that the lateral exit windows are close to the beam entrance window and the cell center. This minimizes the absorption before the photoluminescence detection zone, thus maximizing the luminescence signal intensity.

Components

For the absorption and photoluminescence measurements of diamondoids the multi-purpose cell needs to be heatable, vacuum-tight, and transparent into the vacuum ultraviolet (VUV) spectral region. Further, it is designed to facilitate the sequential execution of the required experimental steps, such as outgassing of the sample or baking out the cell. To complete the gas cell the cell body is equipped with several additional components:

The **windows** are the crucial component of the gas cell because they have to combine several indispensable properties, such as high thermal stability, optical

4.2. MULTI-PURPOSE GAS CELL

Figure 4.2: Picture of the current gas cell setup for absorption / photoluminescence maesurements.

transparency into the VUV spectral region, and vacuum tightness. No commercially available window is able to satisfy all the above to the required degree at the same time. Thus, heatable, VUV-transparent, and vacuum-tight windows, schematically shown in Fig. 4.3, have been designed.

To assure tightness the windows are composed of two frame elements which hold the pane squeezed between two o-rings, as shown in Fig. 4.3. They are screwed against one another using titanium screws to avoid seizing of the thread due to the heating process. The vacuum-side element (bottom element on the right-hand side of Fig. 4.3) is milled from a CF16 flange and easily connects to standard UHV components.

The high thermal stability is achieved by using Kalrez® o-rings (from DuPont) which are guaranteed to endure temperatures up to 170°C. Even higher temperatures can be achieved temporarily without compromising the functioning of the gas cell.[2] Window panes are made from MgF_2 crystals which are transparent for energies up to about 11 eV which is well beyond the region of interest.[3] At the

---

[2]In our experiments, the windows survived temperatures beyond 200°C. However, around 200°C the O-rings start to outgas giving rise to characteristic spectral signature in the absorption spectra. Also, O-rings that have been heated to 200°C or above become brittle and will in continuing use compromise tightness of the absorption cell.

[3]Here, the ionization potential of adamantane (9.23 eV [25]) provides a useful reference for the upper limit of the energy region.

47

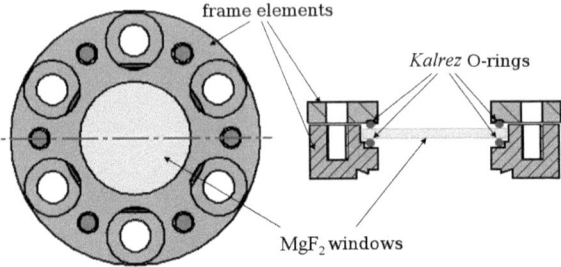

Figure 4.3: Schematic drawing of the constructed MgF$_2$ windows. The pane is held between two frame elements which connect to a regular CF 16 vacuum component. Air tightness is achieved by using Kalrez o-rings which exhibit high thermal stability.

same time MgF$_2$ absorbs infrared radiation with less than 1000 cm^{-1} ($\approx$ 125 meV or 1500 K), as shown in Fig. 4.4, meaning that the windows are efficiently heated and at thermal equilibrium with the rest of the gas cell. This ensures that the diamondoid samples do not condense on the windows. This is important to make sure that the acquired data indeed stems from isolated particles in the gas phase rather then conglomerates of condensed diamondoid crystals. Several experimental runs showed that indeed samples do not condense on the windows, but in the case of supersaturation[4] which can be excluded for all of the presented data.

A **valve** is installed on the top of the absorption cell. It allows to load the cell without dissembling any fixed parts. It can be opened and closed within the vacuum chamber from the outside through a mechanical feed-through, as indicated in Fig. 4.7. Being able to open and close the valve within the cell serves multiple purposes. It allows to evacuate and then close the cell before starting to heat and sublimate the sample and also to open the cell for the bake out after the measurement. Further, measuring the transmission of the open and the supposedly empty closed cell provides a way to confirm a complete and successful bake out.

---

[4] At one occasion diamantane condensed on the cell windows after filling the cell at elevated temperatures (T= 79°) and then letting the closed cell cool down. Consequently, heed was taken to not fill the cell in a fast cooling process and the condensation of samples on the windows can be ruled out for all of the discussed data.

4.2. MULTI-PURPOSE GAS CELL

Heating wires are used to allow **heating** the gas cell to temperatures providing sufficient vapor pressures of the contained samples. The heating wires are tightly wrapped around the cell and at more solid parts, such as the valve and flanges, fixed with clamps to ensure a sound thermal contact. If high temperatures ($> 150°C$) are required the cell can be partially or fully wrapped in aluminum foil which functions as a heat shield and thus maximizes the heating efficiency. While this allows faster and more efficient heating it also entails a slower cooling process, especially in vacuum where large parts of the heat loss occur by means of radiation. For samples which are available in very small amounts fast heating is advantageous because it minimizes the time until the measuring temperature is reached and, thus, the small yet non-negligible sample loss through leakage.

The heating process is steered directly via the temperature of the heating wire. A thermocontroller using a standard thermocouple reads out the temperature of the heating wire and compares it to the set temperature. A second thermocouple is connected to a comparably cold part of the cell to monitor the progress of the heating process. This setup allows to control minimum and maximum temperatures of the cell and helps to avoid large temperature gradients which would induce mechanical strain.

For all connections titanium or silver-plated screws are used to avoid a seizure of the thread upon heating. The use of vacuum lubricants was avoided to minimize the risk of contaminations within the gas cell.

Ceramic spacers were used to keep the thermocouples and the wire leads of the heating wires from making electric contact with one another or with the chamber walls. All wires are lead to the outside of the vacuum chamber through electrical feedthroughs.

The gas cell is connected to a flange which is fixed on the side of the chamber. The flange is a special construction and can be moved around by several millimeters to adjust the cell position in up-down and in the beam direction. The cell is held by an assembly of four threaded 5 mm metal rods. The holding construction allows for the adjustment of the cell in left-right direction (from the beam). It is designed to combine minimal thermal contact to the external vacuum chamber with the rigid stability which is necessary to cope with the forces when opening and tightly closing the valve of the cell.

CHAPTER 4. EXPERIMENTAL METHODS

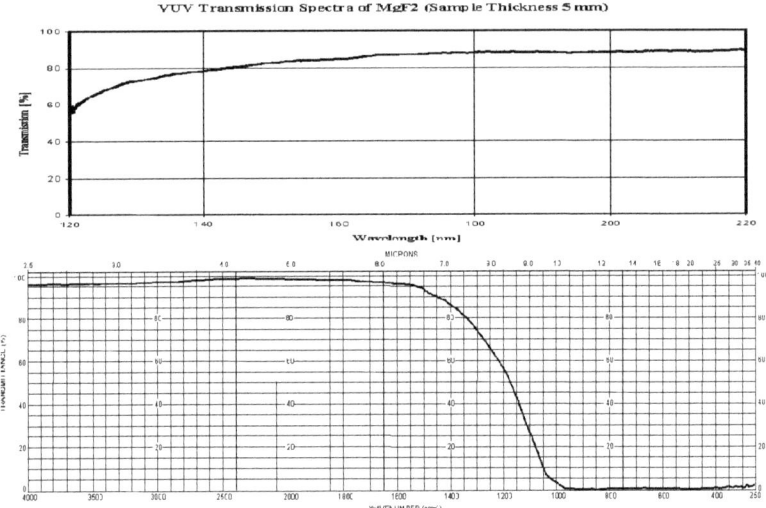

Figure 4.4: The transmission curves of the MgF$_2$ windows [80] in the ultraviolet (top) and infrared (bottom) spectral region.

Operation

For absorption and / or photoluminescence measurements the gas cell is operated as follows:

Outside the vacuum chamber, the gas cell is filled through the open valve with a few mg of the sample. The absorption cell on the holding flange (compare Figs. 4.1 and 4.2) is then put into place and vacuum chamber is pumped down. The cell is opened shortly to allow the residual air to escape. Where the samples require heating, the cell is slowly heated to avoid strain. The sample is outgassed at ambient pressures smaller than $3 \cdot 10^{-5}$ mbar. The temperature is stabilized when a sufficient vapor pressure of the sample is reached. Measuring temperatures for all diamondoids are listed in Tab. 4.1. The absorption signal is used as a rough indicator for the vapor pressure and measurements are conducted at an estimated vapor pressure on the order of ~ 0.1 mbar.

After the measurement of a sample is complete the valve of the absorption cell is opened completely and the cell is baked out over an extended period of time (typically 30 min to 1 hour) at a temperature exceeding the measuring temperature. The success of the bake out is controlled by scanning for an absorption signal of the closed cell. After the bake out is complete the cell is cooled down and the chamber is vented. For the next sample the same procedure is repeated.

4.3 Synchrotron Radiation

For optical absorption as well as photoluminescence measurements, which are described in sections 4.4 and 4.5, respectively, synchrotron radiation was used as the light source.

Synchrotron radiation results from the relativistic dipole radiation which is emitted by accelerated charged particles moving at velocities close to the speed of light. Typically, electrons which move at more than $0.999\,c$ are accelerated in a horizontal plane, i.e., orthogonally to their direction of motion, by a periodic magnetic field, a so-called insertion device. Due to relativistic effects the dipole radiation is strongly concentrated in forward direction. In modern 3rd generation synchrotron facilities, such as BESSY II, typically an undulator is used as insertion device. The principle of undulator radiation generated in a synchrotron is shown in Fig. 4.5. In an undulator the horizontal displacement of the electron bunches in the magnetic field is small enough for the lobes of the individual oscillations to interfere with each other [82]. The parameters of the alternating magnetic field and the electron bunches are matched such that the interference at the desired energy occurs constructively. Due to the constructive interference the intensity of the beam grows with the square of the number of oscillations in the magnetic field, i.e., quadratic with N, the number of pairs of dipole magnets. The resulting undulator peak consists of an energy range which is typically too broad for spectroscopy purposes. Therefore a monochromator is employed to select a narrow wavelength region of the undulator radiation. To scan the energy the undulator gap and the monochromator grating are adjusted synchronously and stepwise to yield a maximum intensity at the desired wavelength. The energy resolution is determined by the monochromator but can be varied by opening or closing the exit slit which comes at gain or loss of beam intensity, respectively.

CHAPTER 4. EXPERIMENTAL METHODS

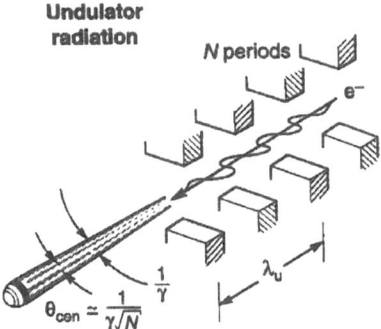

Figure 4.5: The principle of undulator radiation. Electrons which move at velocities near the speed of light undergo an oscillatory motion in a periodic magnetic field. Due to relativistic effects the resulting dipole radiation is strongly collimated in forward direction. In an undulator the lobes resulting from the single oscillations interfere constructively, thus the intensity of the beam grows as N^2. [81]

4.4 Optical Absorption Spectroscopy

Optical absorption spectroscopy probes band-to-band transitions between the valence and the conduction band, or - depending on the notation - the HOMO and the LUMO, respectively. A schematic view is given in Fig. 4.6. The investigated system is promoted from the electronic ground state into an excited state through the absorption of a photon. The change in photon transmission is measured as a function of photon energy. The absorption signal is a measure for the transition probability between all pairs of filled and empty electronic states whose energy difference matches the photon energy. For some transitions which are dipole-forbidden (compare section 3.3) this probability is zero. For semiconductors, the absorption of photons does not occur below an energy threshold. This energy is usually

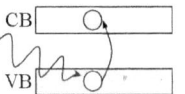

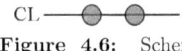

Figure 4.6: Schematic drawing of the principle of optical absorption spectroscopy. CB, VB, and CL label the conduction band, the valence band and the core-levels, respectively.

4.4. OPTICAL ABSORPTION SPECTROSCOPY

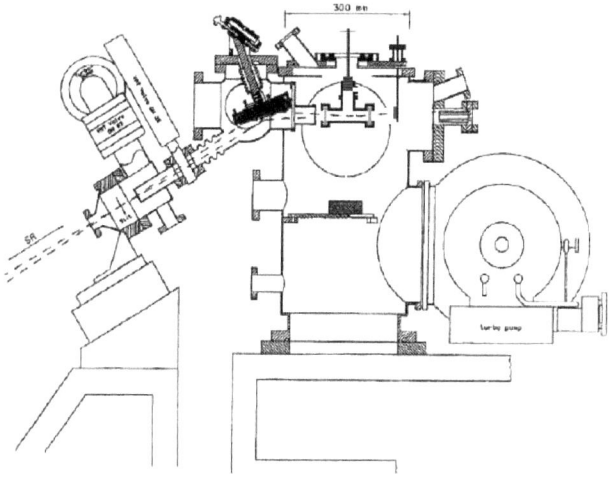

Figure 4.7: The experimental setup for the absorption measurements at the Hasylab. The schematic drawing shows a lateral cut through the Clulu vacuum chamber. The synchrotron beam arrives from the left at an angle and is reflected into the main chamber. It traverses the gas cell and hits a photodiode which measures the transmitted intensity of the beam. [83]

labeled as *gap energy*, E_{gap}.[5]

Experimental Setup

The optical absorption of the diamond clusters is determined as difference of the measured transmission between a filled and an empty absorption cell. The absorption of pristine and functionalized diamondoids is measured using synchrotron radiation and the gas cell setup described above. The experiments were performed using synchrotron radiation in the vacuum ultraviolet spectral regime at beamline I of Hasylab (DESY, Hamburg) and at beamline U125/2-NIM of the Berlin synchrotron facility BESSY II.

A profile of the setup at Hasylab is shown schematically in Fig. 4.7. The absorption cell is centered within the vacuum chamber such that the synchrotron light traverses the cell before hitting a photodiode at the far end of the chamber. The

---
[5]The precise definition of the energy gap is far from being unambiguous. Further differentiation will be undertaken in section 5.1.3.

CHAPTER 4. EXPERIMENTAL METHODS

diode current is read out using a Keithley picoampèremeter. The intensity of the transmitted light is then recorded as function of the monochromator position. The difference of two transmission scans for the empty and the filled absorption cell provides the absorption signal of a sample. The pressure curves for most of the measured samples are unknown and thus the relative not the total absorption cross section has been determined[6].

Some of the absorption measurements, e.g. on oxadiamondoids, were conducted at BESSY II and already used the advanced version of the gas cell shown in Fig. 4.2 which allowed for simultaneous detection of photoluminescence. The principle, however, stays the same and the altered gas cell setup did not affect the outcome of absorption measurements in any way.

Data Acquisition and Analysis

The transmission signal is measured as the current of a photodiode behind the gas cell which is read out by a Keithley picoampèremeter. During a scan over the investigated energy range the diode current is written to a file together with the monochromator position. For each sample a scan of the empty gas cell has been recorded directly before or after the scan of the sample. This ensures that a possible degradation of the transmission properties of the windows as well as potentially persistent contaminations on the window surface are accounted for.

The two scans of the empty and the filled cell are scaled to account for differences in the ring current of the synchrotron. The absorption cross section can then be deduced from the transmitted intensities, I_0 and I, according to the Lambert-Beer law:

$$\frac{I}{I_0} = e^{-\sigma \ell N} \tag{4.1}$$

or

$$\sigma = -\frac{1}{\ell N} ln\left(\frac{I}{I_0}\right). \tag{4.2}$$

The exact density of the particles N inside the gas cell is unknown because the vapor pressure as a function of temperature has not yet been determined. The temperature is kept constant throughout each scan and the determined absorption

---
[6]Note that pressure gauges used in the sub-mbar regime are typically calibrated to N_2 and correction coefficients exist for many other gases. For diamondoids the correction coefficients have not yet been determined and pressure gauge readouts could easily be off by an order of magnitude or more.

4.4. OPTICAL ABSORPTION SPECTROSCOPY

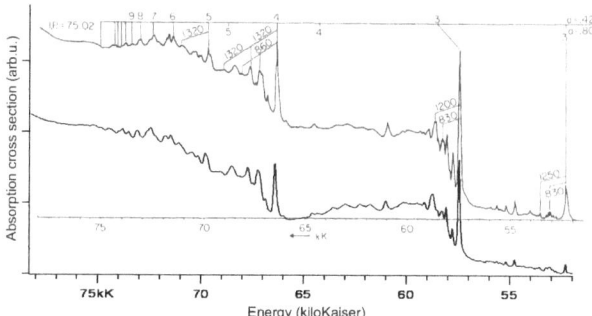

Figure 4.8: Comparison of the adamantane spectrum recorded using the experimental setup described in section 4.2 (bottom) with an adamantane from the literature (top) [32]. All features are reproduced showing that the experimental data are not limited by the spectral resolution. The energy scale is given in kiloKaiser (1kK = 1000 cm^{-1} = 124 meV), corresponding to the reference.

cross sections are relative cross sections. The spectra can therefore be discussed in terms of the relative intensities of spectral features.

For the pristine diamondoids the spectra have been recorded in steps of 0.1 nm and the spectral resolution is better than 20 meV over the entire spectral range. For the diamondoid thiols a step width of 0.2 nm was used. The resulting resolution of the spectra is still better than 30 meV and is sufficient to spectrally resolve the comparably smooth spectra (compare section 6.2). For all other samples the step width of the monochromator has been adjusted throughout the scan by defining different step sizes for different spectral regions. Therefore, the nominal spectral resolution varies throughout the spectrum. The step width has been set to yield a resolution better than 20 meV in energy regions where spectral features are abundant.

Proof-of-Principle

In Fig. 4.8 the adamantane spectrum which has been recorded at the Hasylab using the gas cell setup described in section 4.2 is compared to a reference spectrum from the literature [32]. The two curves show excellent agreement. Further, the spectrum recorded in the present experimental setup reproduces all of the spectral features found in the reference spectrum which has been recorded with a resolution better than 3 meV [32].

This comparison demonstrates the suitability of the absorption spectroscopy setup and shows that the spectral resolution is sufficient to resolve all spectral features. Because, of all investigated samples, the adamantane spectrum exhibits the sharpest spectral features, it follows that the resolution does not pose a limit on any of the optical absorption spectra presented in this work.

4.5 Photoluminescence Spectroscopy

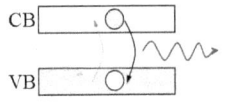

Figure 4.9: Schematic drawing of the principle of photoluminescence spectroscopy.

Just like absorption spectroscopy, photoluminescence spectroscopy probes the band-to-band transitions of a system. The system, in our case the diamondoid, is promoted to an electronically excited state and emits a photon when it relaxes back into its electronic ground state, as sketched in Fig. 4.9. This photon carries important information about the system which is supplemental to the information acquired using absorption spectroscopy. In particular, information about the excited state, such as the magnitude of the geometric distortion, can be retrieved.

The idea to measure photoluminescence of diamondoids came from drawing analogies to Si-nanostructures. Silicon, like diamond, is an indirect band gap semiconductor and, therefore, in its pure state not prone to emit light upon excitation. However, it has been observed that Si-nanostructures exhibit enhanced luminescence [69]. This has been explained by a breakdown of the momentum conservation rule in nanostructures [68], a concept that should be universal in its nature and, therefore, also apply to the indirect semiconductor diamond.

Experimental Setup

The photoluminescence of diamondoids was measured at the NIM endstation of beamline U125/2 at the Berlin synchrotron facility BESSY II. The incoming photon beam is monochromatized using a 10 m normal-incidence monochromator

4.5. Photoluminescence Spectroscopy

equipped with a 300 l/mm grating. To confine the interaction of the synchrotron beam with the sample to a small, well-defined volume, the gas cell setup described in section 4.2 is used. The cell is placed inside a vacuum chamber such that it is traversed by the synchrotron beam. Behind the absorption cell the transmitted light is detected by a photodiode, similar to the experimental setup for the absorption measurements. The light transmitted through the gas cell provides measure for the absorbed photons which is also used in section 5.2 to estimate the quantum yield. The enhanced experimental setup is used for the simultaneous measurements of absorption and time- and spectrally resolved photoluminescence. The photoluminescence signal is detected on either side of the vacuum chamber at an angle of 90° to the incident beam. On one side the setup features a photomultiplier to deliver the time-resolved luminescence signal[7]. On the other side of the current multi-purpose setup the luminescence light is dispersed by a 1 m normal-incidence monochromator equipped with a 1200 lines/mm grating (blaze wavelength 150 nm) and recorded by a position-sensitive CsTe microchannel-plate detector. The detector is sensitive to wavelengths from 300 nm down to 160 nm [85] and has a fairly constant quantum efficiency down to 200 nm which decreases slightly towards 160 nm [85, 86]. The detector, provided by the group of A. Ehresmann (U Kassel), has never been gauged for absolute quantum efficiencies [86]. However, over the relevant energy range the CsTe-detector exhibits a spectral response that does demand corrections of the spectra in their relative intensities [85].

The photoluminescence setup allows for the detection of spectrally resolved data as well as of the total luminescence yield. For the latter the secondary monochromator is operated in 0-order, effectively integrating the luminescence signal over the entire spectral range. The measurement of the total luminescence yield as a function of excitation energy is known as luminescence excitation and also provides a direct measure for the light absorption.

The (primary) monochromator of the synchrotron beamline yields a maximum resolution of better than 30 meV for slit settings of 100 μm for entrance and exit slit. For the recording of spectrally resolved luminescence of diamondoids the exit slit was opened to 400 μm. This increased the photon flux fourfold at the cost of the resolution of the synchrotron beam. An effect of an energetically broader excitation on the recorded luminescence spectra was not found.

---

[7]As mentioned earlier, the results of the time-resolved luminescence measurements will be discussed in a later work [84].

CHAPTER 4. EXPERIMENTAL METHODS

The spectral resolution of the monochromator-detector combination for the detection of diamondoid photoluminescence has been determined to be better than 30 meV.

To suppress stray and scattered light as well as signal contributions from the background a pinhole at the front window of the cell and an exit slit on the luminescence window were installed. The 3 mm pinhole before the cell diminished the effect of reflections from the beamline. The exit slit, which is in its final version rather a combination of slits along the path of photoluminescence signal as shown in the picture in Fig. 4.2, proved useful to suppress different troublesome signal spots on the detector image which appeared to stem from reflections within the absorption cell. By reducing the area of the window by a factor of twelve the slit also reduced the background noise from fluorescence light reflected within the absorption cell. As an adverse effect the slit also leads to a reduction of the solid angle of the detection area by approximately a factor of three. Blackening the inside of the absorption cell (to UV light) might make the use of a slit dispensable. Experience has shown, however, that tiny desorption rates can lead to notable contaminations within the unpumped volume of the gas cell. Typical tools, such as graphite spray, do not provide the vacuum compatibility needed to be used in such an unpumped, heated vacuum cell.

Data Acquisition and Analysis

Different kinds of measurements are possible for each diamondoid sample. To derive the spectra the following steps have been taken:

For a **luminescence excitation spectrum** the excitation energy is scanned over the corresponding energy region, typically in steps of 0.2 nm. The CsTe detector is triggered by the step motor of the beamline monochromator and the total luminescence signal for each energy step is recorded for 5 seconds. A program writes the step number and the signal intensity to a text file which, together with the settings of the beamline monochromator, allows the reconstruction of the luminescence excitation spectrum.

For the recording of a **photoluminescence spectrum** the beamline monochromator is set to a constant excitation energy. The excitation energies are chosen to excite the diamondoids in a strong absorption band to yield a high signal

4.5. PHOTOLUMINESCENCE SPECTROSCOPY

Diamondoid name	T (K)	E_{exc} (eV)	E^*_{gap} (eV)
adamantane	293	7.12	6.49
diamantane	360-375	7.90	6.40
triamantane	389-397	7.43	6.06
[121] tetramantane	410-438	7.75	6.10
[123] tetramantane	411-414	7.90	5.95
[1(2)3] tetramantane	408-418	7.75	5.94
[1212] pentamantane	420-421	7.87	5.85
[1(2,3)4] pentamantane	407-409	7.75 (7.20)	5.81

Table 4.1: Temperatures and excitation energies at which the photoluminescence of each diamondoid has been measured. For [1(2,3)4] pentamantane a second spectrum has been recorded at a different excitation energy to check for a dependence of fluorescence lines on the excitation wavelength. The photoluminescence spectra are discussed in section 5.2. The band gap is also listed for comparison. (*More precisely, this is the optical gap as defined later in section 5.1.3)

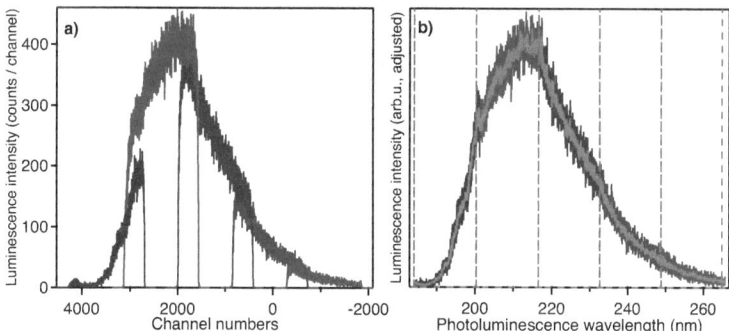

Figure 4.10: Construction of a photoluminescence spectrum from multiple detector images (partial spectra) using the example of adamantane. In the left panel the original data are shown. In the right panel they have been scaled to the same height at the merging points (dashed lines). To merge the spectra with highest possible accuracy each of the individual spectral fragments has been smoothed and the final spectrum has been adjusted according to the smoothed partial spectra (right panel).

CHAPTER 4. EXPERIMENTAL METHODS

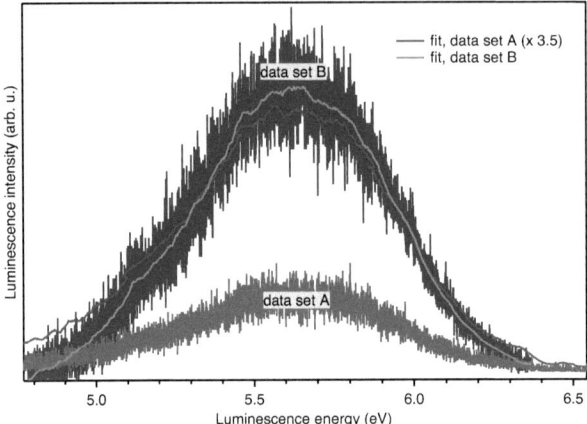

Figure 4.11: The graph shows two different sets of photoluminescence data for diamantane. Data set A has been recorded for fewer time but under the same circumstances as data set B. Both spectra have been constructed according to the description in the text. The fit for data set A has been scaled in height to match the fit of data set B. It can be seen that the agreement is fairly well.

intensity. They are listed in Tab. 4.1. The resulting photoluminescence is then detected by the spatially resolving CsTe multichannel plate-based detector.

Prior to the first measurements the secondary monochromator has been calibrated using the reflections of the synchrotron beam on a solid, largely transparent medium which was placed inside the absorption cell for this purpose. The primary monochromator was moved from 175 nm to 270 nm in steps of 5 nm and the position of the signal and the grating position of the secondary monochromator were noted. This allowed to precisely calibrate the setup over the entire spectral range.

The photoluminescence spectra of diamondoids turned out to be unexpectedly broad in energy. Distinct sharp features in the spectra are absent. Spectra typically stretch over several tens of nanometers, or more than 1 eV, as is shown in Fig. 4.10 and as will be discussed in detail in section 5.2. As a consequence, spectrally resolved photoluminescence spectra had to be recorded in a sequence of multiple recordings at different detection energies. The resulting partial spectra were afterwards merged to a single photoluminescence spectrum. This procedure is exemplified in Fig. 4.10 using the example of adamantane. In the case of

4.5. PHOTOLUMINESCENCE SPECTROSCOPY

adamantane five spectra were recorded at a different position of the grating, i.e., at different wavelengths. Later, for larger diamondoids, the individual grating positions were further optimized such that the number of partial spectra could be reduced to four. The recorded wavelength regions were chosen to exhibit a broad spectral overlap which allows to afterwards adjust and connect the partial spectra to one another. Panel a) of Fig. 4.10 shows the original data for adamantane over the channel numbers which have been adjusted for the different grating positions. Each partial spectrum has been recorded for 30 min. The fact that at the edges the partial spectra do not perfectly match in height is due to the finite lifetime of the electron bunches in the storage ring which results in a decreasing photon flux over time. In some cases additional effects arise from small variances in the cell temperature between different partial scans (comp. Tab. 4.1) resulting in different vapor pressures and, consequently, differences in absorption / emission intensities. For each partial spectrum an individual fit was created by smoothing. The smoothed lines, shown in Fig. 4.10 b), were then used to scale and fit the spectra to one another. As a last step, overlapping region of the partial spectra were cut along the dashed lines and joint to produce a single, complete luminescence spectrum.

The recorded photoluminescence signal was fairly noisy. This impression is aggravated by the high resolution of the detector which distributes the entire photoluminescence peak over a total of more than 4000 channels for each spectrum.

The reconstruction procedure of the photoluminescence spectra from multiple scans and the signal-to-noise ratio raises the question of reproducibility of the spectra. In Fig. 4.11 two different measurements of the photoluminescence spectrum of diamantane are shown. Partial spectra of both data sets have been recorded for only 30 min. Data set A, however, has been recorded with a much lower beamline flux than data set B. All other experimental parameters, such as the excitation energy etc., were kept the same. Therefore data set B, which is the actual data set used in the following discussion, has much better statistics. However, when scaling the fit of data set A to similar height it becomes obvious that the two scans are largely equivalent. The worse signal-to-noise ratio caused by the lower signal intensity leads to an apparent broadening of the photoluminescence peak for data set A. The energetic position as well as the shape of the peaks are nevertheless in good agreement despite the very low statistics of data set A. This demonstrates that the photoluminescence data acquired during this thesis are reproducible and allow for a meaningful discussion of the spectra and

CHAPTER 4. EXPERIMENTAL METHODS

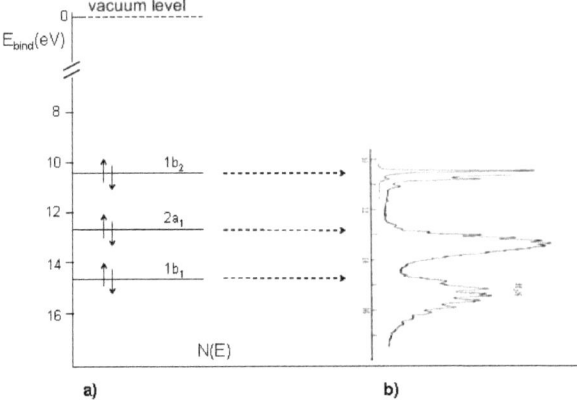

Figure 4.12: The assignment of electronic levels in a photoelectron spectrum using the example of H_2S [87]. Depending on the intensity with which electronic levels couple to vibrations the photoelectron peaks turn out either sharp (low coupling, top) or broad (strong coupling, bottom).

even single features. The implications of the low statistics on the possibilities of a qualitative and a quantitative discussion of the spectral details are discussed with the results in section 5.2.

4.6 Ultraviolet Photoelectron Spectroscopy

Photoelectron spectroscopy is an experimental tool to determine the binding energies of electrons in solids or molecular systems by making use of the photoelectric effect: Electrons are excited from the investigated system into the continuum using monochromatized light ($E_{ph} = h\nu$), schematically shown in Fig. 4.13. The kinetic energy of the photoelectrons is the difference of the photon energy and their binding energy and thus the binding energy is given by

$$E_{bind} = h\nu - E_{kin}. \qquad (4.3)$$

By scanning the kinetic energy of the photoelectrons one acquires a spectrum that corresponds to the electronic levels of the investigated system.
Fig. 4.12 visualizes the interdependence of a photoelectron spectrum and the electronic levels of the investigated system. The electronic levels of polyatomic, more complex structures typically do not appear as sharp lines but as broad peaks or

4.6. ULTRAVIOLET PHOTOELECTRON SPECTROSCOPY

even bands due to the excitation of vibrational modes which give rise to so-called *vibronic progressions* and the proximity of electronic levels.

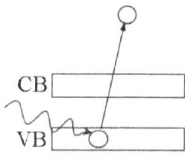

Photoelectron spectroscopy comes in two flavors: X-ray photoelectron spectroscopy (XPS) and valence band photoelectron spectroscopy which can be performed using synchrotron radiation (oftentimes simply referred to as photoelectron spectroscopy, PES) or an ultraviolet laboratory-based light source (UPS). All three methods have been applied to pristine and/or functionalized diamondoids.

Figure 4.13: Principle of ultraviolet photoelectron spectroscopy.

Within the frame of this thesis ultraviolet photoelectron spectroscopy (UPS) has been employed to probe the highest occupied states of thiolated diamondoids. For the XPS and PES measurements on pristine diamondoids, which were also conducted in our group, only parts of the interpretation of the data have been accomplished within this thesis and a summary of the results will be published in Ref. [37]. The entire experimental work and large parts of the data analysis have been carried out by Kathrin Klünder and can be found in detail in her diploma thesis [26].

Experimental Setup

The UPS setup used an oven to evaporate the samples into a vacuum chamber which features a photoelectron spectrometer and a helium lamp as light source.

The diamondoid samples are placed inside an oven and evaporated as a particle beam through a nozzle into the vacuum chamber. The oven, shown in Fig. 4.14, possesses a little pot (inset of Fig. 4.14) which holds the sample and is wrapped by a heating wire. By controlling the temperature of the sample pot via a thermocontroller and a thermocouple connected to the lid of the sample pot the vapor pressure of the sample and thus the density of the particle beam leaving the nozzle can be controlled.

The oven is adjusted such, that the particle beam is crossed with the light of a He-lamp under the aperture of the electron analyzer. The He-lamp (Leybold-Heraeus, type UVS 10/35) provides intense and monochromatic ultraviolet light with a photon energy 21.22 eV. This photon energy of the He(I)-line clearly ex-

CHAPTER 4. EXPERIMENTAL METHODS

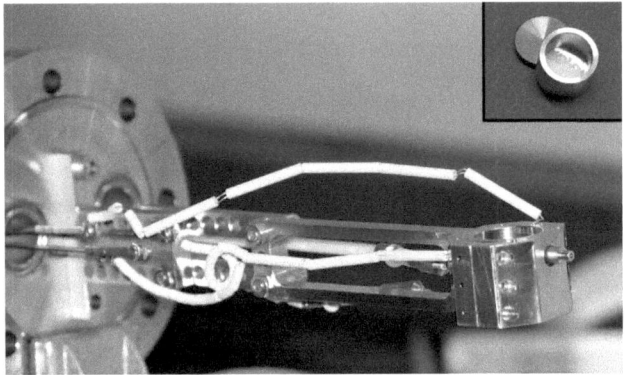

Figure 4.14: Photograph of the oven that is used in the PES setup. The sample is placed inside the pot, shown in the inset, which is placed inside the oven and wrapped by a heating wire. The oven temperature is monitored by two thermocouple shown in the picture. A copper nozzle is used to avoid clogging. [26]

ceeds the ionization potentials of diamondoids ($< 10\,\text{eV}$) and therefore provides sufficient photon energy to record valence band spectra.
As electron analyzer a Scienta SES-2002 is used. The interaction region is shielded from outside influences by a μ-metal pipe. The detector uses two hemispheres at different potentials to adjust the pass energy at which the photoelectrons reach the detector. A pair of multi-channel plates (MCPs) amplifies the electron signal which is then made visible on a phosphorescent screen. The screen is filmed by a CCD-camera and the image is read out by a computer software. By scanning the kinetic energy over the desired energy range a spectrum is acquired which can, with the help of equation 4.3, be converted into a spectrum of the valence level binding energies.

Data Acquisition and Analysis

Prior to each run the sample pot and the nozzle are cleaned with acetone in an ultrasonic bath to ensure that no residues or contaminations are present. The new sample is loaded and the oven is installed in the chamber. After the He-lamp has been ignited the Scienta detector is switched on and the region of interest for kinetic energy of the electrons is scanned. The oven is heated until notable

4.6. ULTRAVIOLET PHOTOELECTRON SPECTROSCOPY

Sample	T_{wire} [°C]	T_{pot} [°C]	$p_{chamber}$ [10^{-6}mbar]	Experimental Resolution [meV]
adamantane-1-thiol	-	23	5,2	40
diamantane-4-thiol	70-75	49-58	5,5	40
diamantane-1-thiol	100-120	64-86	4,7	40
triamantane-9-thiol	140-150	75-91	1,9	50
triamantane-3-thiol	120-140	85-119	4,5	60
[121] tetramantane-2-thiol	140-150	92-105	1,7	80
[121] tetramantane-6,13-dithiol	220	140-153	1,7	80
[1(2,3)4] pentamantane-7-thiol	170-190	109-132	1,0	80

Table 4.2: This table lists the most important experimental parameters for the UPS measurements of the diamondoid thiols. T_{wire} and T_{pot} give the temperatures of the heating wire and the sample pot, respectively. $p_{chamber}$ is the background pressure in the vacuum chamber during the measurement. In the last column the spectral resolution is given for each of the measured samples.

increase in the chamber pressure, $p_{chamber}$, is noted and a photoelectron signal is detected. An energy region with a width of $4-5$ eV is scanned in several sweeps. One sweep of the energy region takes approximately 1 min which allows to take spectra even during a slow heating process (compare temperatures in table 4.2). The error in assuming a constant particle density during a sweep is negligible because the changes in temperature listed in table 4.2 occur over a much longer time period of $\sim$1 hour. The sweeps are added to give a spectrum of the kinetic energies of the photoelectrons. Using equation 4.3 the spectra are converted to plot the photoelectron signal over the binding energy.

A Krypton reference spectrum has been taken before each sample to calibrate the spectrometer and to determine the spectral resolution for the applied settings. For the calibration the well known $4p_{3/2}$ line of Krypton at a binding energy of 13.9996 eV [88] is used. The spectral resolution for each scan is listed in table 4.2. For larger diamondoid thiols smaller sample amounts were available and, thus, settings were optimized for a faster scan at the cost of a lower spectral resolution.

Chapter 5

Results & Discussion - Part I: Pristine Diamondoids

In this first part of the "Results & Discussion" segment of the thesis the data on the optical properties of pristine diamondoids are presented. The absorption and photoluminescence data provide information about the influence of nanocrystal size and shape on the optical properties in general and the optical gap in particular. As one of the defining materials' properties the band gap is of particular interest and will be the central matter of discussion in section 5.1.3.

The experimental results presented in this chapter are the first to determine the *exact* influence of size and shape on the optical response of a semiconductor nanocrystal; *exact* meaning that structures are known with absolute, atomical precision and that changes in the optical properties can be directly linked to well-defined changes in size and/or shape of the nanocrystal.

The optical data in this chapter complement the previous electronic structure investigations presented in section 2.3 to render a comprehensive picture of the electronic properties of a series of size- and shape-selected diamond nanocrystals. Due to its perfect structural definition this series can be regarded as a model system for group IV nanocrystals in particular and for small semiconductor nanocrystals in general. Further, all experiments in this series have been designed to provide data that come as close to theoretical purity as experimentally possible. This allows for a direct comparison to typical quantum chemical calculations. Such a comparison to existing theoretical (and experimental) investigations will be given in the final section of this chapter.

CHAPTER 5. RESULTS & DISCUSSION - PART I: PRISTINE DIAMONDOIDS

5.1 Optical Absorption

The optical absorption of diamondoids ranging in size from adamantane to cyclohexamantane has been measured. In Fig. 5.1 the absorption data for eleven distinct diamondoid structures are shown. To the left side of the graph the diamondoid structures are displayed next to the corresponding spectrum. The bulk diamond band gap is indicated by a vertical dashed line at an energy of 5.47 eV [89]. Strong differences in the optical response among different diamondoid structures are evident.

A subset of the spectra has been recorded during my diploma thesis but the discussion and understanding was limited to single spectral features and their size dependent shifts [79]. In this thesis systematic trends within the evolution of the spectra are worked out and a definition of the optical gap is given which allows a detailed comparison to theoretical predictions. The discussion of spectral features from my diploma thesis is not repeated in detail.

First, in section 5.1.1, the optical response of diamondoids is analyzed and a categorization of diamondoids in 1-, 2-, and 3-dimensional diamondoid structures is proposed. In section 5.1.3 the energy gap will be determined for all diamondoids. The different factors which influence the optical gap will be analyzed and the results will be discussed in the light of theoretical and experimental work on diamondoids and similar systems.

A summary of the findings on the optical absorption of pristine diamondoids presented in this section was published under the title *Optical Response of Diamond Nanocrystals as a Function of Particle Size, Shape, and Symmetry* in 2009 in *Physical Review Letters*.[1]

5.1.1 Nanodiamonds in 1D, 2D, and 3D

The optical response of diamondoids, shown in Fig. 5.1, varies greatly among the different diamondoids. It changes not only with the size of the diamondoid but also with its shape, as can be seen, e.g., for the four different pentamantane spectra (2nd to 5th spectrum from the top). The optical response, therefore, obviously strongly depends on the very structure of the individual diamondoid.

---

[1]L. Landt et al., *Optical Response of Diamond Nanocrystals as a Function of Particle Size, Shape, and Symmetry*, Physical Review Letters **103**, 047402 (2009) - Ref. [90]

5.1. Optical Absorption

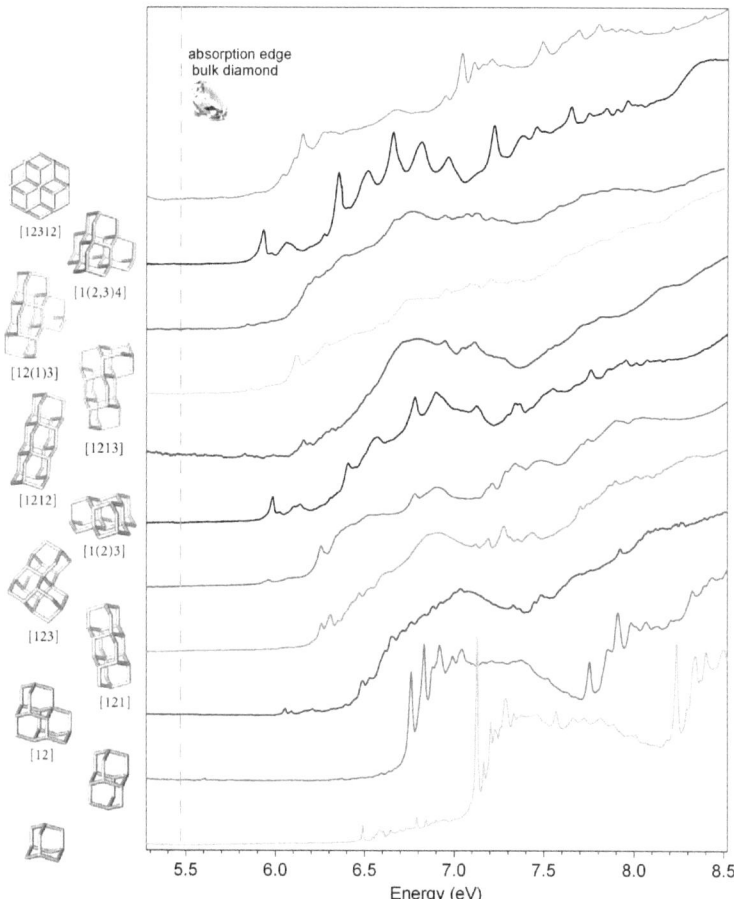

Figure 5.1: Optical absorption of the series of investigated diamondoids. The diamondoid structures are displayed on the left next to the corresponding graph. A dashed line indicates the band gap of bulk diamond [89].

In the following, a categorization scheme of diamondoids according to both their structural and spectral properties is introduced. The categorization scheme, which is shown for the diamondoid structures in Fig. 5.2, is organized as a matrix of diamondoid size and shape. In Fig. 5.3 the spectra are arranged correspondingly. This matrix approach allows to easily distinguish size from shape effects. The categories of this matrix approach, size (vertical) and shape (horizontal), are applicable to higher diamondoids ($\geq$tetramantane) in a strict sense. Lower diamondoids (<tetramantane) form a category of their own which does not obey the matrix scheme. However, there are good reasons to group lower diamondoids, as explained below, even though this does not come about with the same obviousness. The scheme is further arranged as a *growth scheme* which is based upon the idea that the next larger diamondoid can be derived from a given diamondoid structure by adding another carbon cage in a suitable way. In Fig. 5.2 the addition of a diamondoid cage unit is indicated by an arrow between two structures.

Starting point for all diamondoid structures is adamantane, the smallest of the diamondoids. From it diamantane, and successively triamantane, can be constructed through the addition of another diamondoid cage unit as indicated in Fig. 5.2. The spectra of adamantane and diamantane exhibit several very sharp features which are not found in a similarly high intensity and small spacing in any of the other diamondoids' spectra. A sudden break in the spectral appearance occurs when a third diamondoid cage is added. As seen in panel a) of Fig. 5.3 the sharp resonances which, for adamantane, are ascribed to Rydberg states [32] vanish in triamantane. Because of their comparably small size and their sharp spectral features and despite the spectral transition which already occurs at the size of triamantane these three lower diamondoids are summarized in one category as *molecular structures*.

From the structure of triamantane three different tetramantane structures[2] can be constructed by adding an additional diamondoid cage unit. The resulting tetramantane structures are shown in Fig.5.2 and labeled *transitional structures*. The name is due to the fact that through the addition of another cage unit in a suitable manner the three *basic shapes* are derived. Each of the basic shapes can be unambiguously identified with one the idealized *nanodiamond geometries* shown to the right of Fig.5.2. In fact, the size of 26 carbon atoms is the smallest

---

[2]Strictly speaking, there are four tetramantanes because two enantiomers of [123]tetramantane, the 2D transitional structure, exist. Possible differences between the enantiomers have not been investigated in this work.

5.1. OPTICAL ABSORPTION

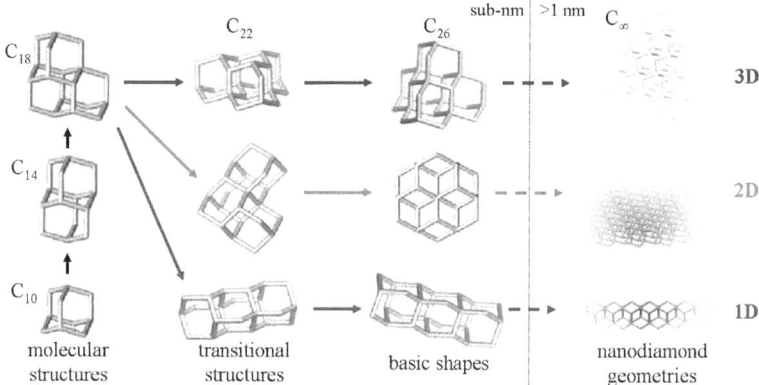

Figure 5.2: Investigated diamondoids categorized by their size and divided into different geometrical families (3D, 2D, 1D). The two middle columns contain clusters of same size. Following the arrows distinguished geometries evolve representing 3D, 2D, 1D nanostructures. Only the carbon framework is shown and the hydrogen surface termination is omitted for clarity.

at which diamond structures can be constructed in recognizably one-, two-, and three-dimensional way. The shown basic shapes, therefore, are the smallest representatives of perfect 1D, 2D and 3D nanodiamond structures, i.e., of diamond nanorods, quantum films and quantum dots, respectively.

The remaining two diamondoid structures, [1213] pentamantane and [12(1)3] pentamantane, like the three basic shapes possess 26 carbon atoms. However, they do not unambiguously belong to a single structural group. Instead, they can be derived from two different tetramantane structures and therefore represent an ambiguous mix of the different geometries.

The distinction between 1D, 2D, and 3D diamondoid structures, which has above been made from a geometry perspective, can also be made from an analysis of the spectra. In fact, it was the observation of spectral resemblances which triggered the thought process that lead to this categorization. In Fig. 5.3 the optical spectra of the diamondoids are arranged in the same way as the diamondoid structures in Fig. 5.2 (without the nanodiamond geometries, of course). The big arrows connect spectra of diamondoids with direct structural links in the same way as in Fig. 5.2. The small arrows indicate the optical gaps which will be discussed in section 5.1.3.

It can be seen that from the smooth spectrum of triamantane in panel a) three

CHAPTER 5. RESULTS & DISCUSSION - PART I: PRISTINE DIAMONDOIDS

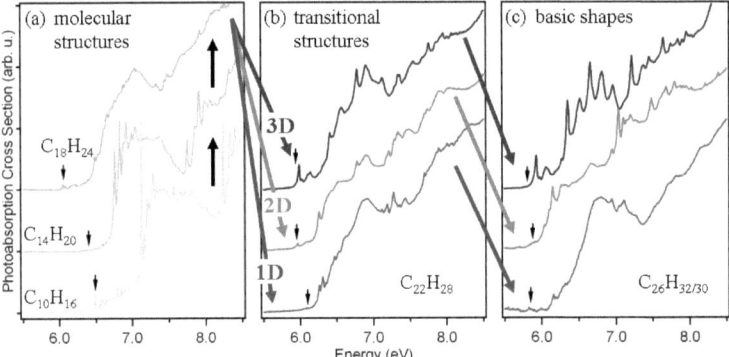

Figure 5.3: Optical absorption as a function of cluster size and shape. The spectra are arranged corresponding to the structures shown in Fig. 5.2. The optical gaps as defined in section 5.1.3 are indicated by small arrows.

very distinct optical spectra evolve for the three transitional structures, i.e., for the three tetramantanes. The spectrum of [1(2)3] tetramantane, as a member of the 3D structural family, is fairly structured and exhibits a strong peak at the absorption onset. In the 2D transitional structure ([123] tetramantane) the absorption onset is far less pronounced and, overall, has less spectral structure. The spectrum of 1D [121] tetramantane only shows sporadic structure. Further it completely lacks the peak at the absorption onset that is observed for the two other structures at energies below 6 eV. Note that all transitional structures have exactly the same size, i.e., the same number of atoms and diamond cage units. The spectral differences therefore demonstrate the drastic influence of the diamondoid shape.

The influence of shape grows with increasing size as can be observed in panel c). The 3D diamondoid possesses by far the strongest resonances among the measured higher diamondoids while the 1D basic shape hardly exhibits any sharp spectral features. Concerning the extent of spectral features the disc-shaped 2D diamondoid is in between the two extremes of the 1D and the 3D basic shape. Yet, it has its own, very distinct spectral signature. Its spectrum exhibits an almost step-like increase of the absorption around 6 and 7 eV, each accompanied by a sharp resonance. Neither one of these resonances matches any peaks of the 3D or 1D basic shapes. A comparison to panel b) shows that the spectral characteristics evolve within each of the structural families: The sharp, regularly

5.1. Optical Absorption

spaced features of the 3D basic shape are already present in a rudimental form in the corresponding transitional structure. Most notably the characteristic double peak at the absorption onset can be recognized. Similarly, the spectral shape evolves for the 2D family where the two characteristic features in the basic shape are also already present in the transitional structure. The 1D family, on the other hand, stands out through a lack of structure altogether.

The above analysis of the optical spectra of the measured higher diamondoids shows that the shape of the diamondoid has a considerable influence on its optical spectrum. The influence of shape on the optical properties even outweighs size effects in many respects. The spectral similarities among diamondoids with a similar shape but a different size are larger than the spectral resemblance of diamondoids of the same size.

Judging only by their geometry [12(1)3] pentamantane and [1213] pentamantane do not fit into the scheme introduced above. From a structural point of view [1213] pentamantane is a chimera of [121] and [123] tetramantane and [12(1)3] pentamantane can even be derived from all three tetramantane structures. They can therefore not be unambiguously assigned to any of the groups by their structure. Nevertheless, the spectra are in the following compared to those of the other diamondoids and it is attempted to assign them to one of the categories from their spectral characteristics. This provides a further test of the plausibility of the linkage between a diamondoid's geometry and its optical response.

Fig. 5.4 shows the spectra of [1213] pentamantane and [12(1)3] pentamantane together with the two other measured pentamantanes. The special role of [1(2,3)4] pentamantane with its many pronounced and regularly spaced peaks which has been pointed out above again becomes evident from this graph. Among the pentamantanes a resemblance of the spectra of [1212] and [1213] pentamantane at the absorption onset is apparent. Both spectra possess peaks around 6.1 and 6.3 eV which are very similar in spectral shape and relative strength. The broad resonance of [1212] pentamantane ranging from approximately 6.3 − 7.3 eV which is characteristic of the 1D structural family does not have a counterpart in the [1213] pentamantane spectrum. Such a broad resonance can, however, be recognized in the spectrum of [12(1)3] pentamantane. Here, the characteristic peak at the absorption onset of [1212] pentamantane around 6.1 eV is not reproduced.

The comparison to diamondoids of the same size yields some insight but no ap-

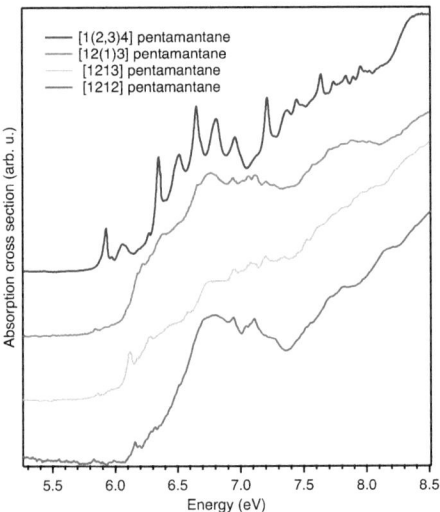

Figure 5.4: A comparison of the absorption spectra of the four different measured pentamantane isomers. All structures have the exact same size and differ only in shape.

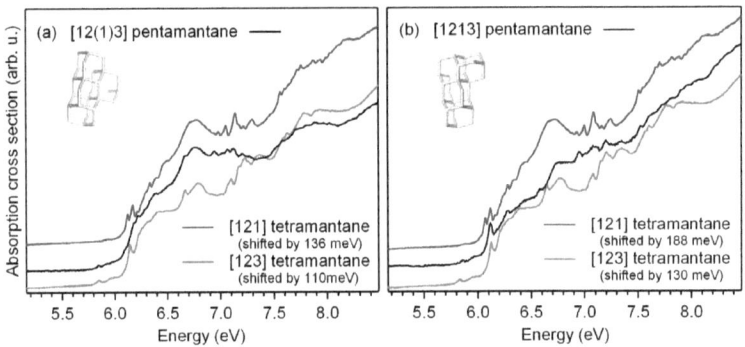

Figure 5.5: Comparison of the spectra of (a) [12(1)3] pentamantane and (b) [1213] pentamantane to those of the tetramantanes. The tetramantane spectra are offset in energy to overlay characteristic spectral structure and along the y-axis for clarity. [1(2)3] tetramantane has not been included because of the lack of spectral similarities with either of the depicted pentamantanes.

5.1. OPTICAL ABSORPTION

parent assignment of either of the structures. To investigate the origin of spectral features it is therefore commendable to compare the spectra to those of the three tetramantanes. This has the advantage that the structures of all larger diamondoids can be derived from (at least) one of the isomers of tetramantane and that the three different isomers exhibit characteristic absorption spectra as seen in Fig. 5.3 (b). In Fig. 5.5 the spectra of the pentamantanes are compared to those of [121] and [123] tetramantane which have been shifted in energy to overlay similar spectral features. [1(2)3] tetramantane is not included because of a palpable lack of spectral resemblance.

The spectrum of the branched [12(1)3] pentamantane in panel (a) does not exhibit any sharp peaks. A very faint feature is observed between 5.8 and 5.9 eV and can be matched to a comparable feature in L-shaped [123] tetramantane. This feature is absent in the I-shaped [121] tetramantane and is therefore tentatively linked to the existence of a bend in the diamondoid structure. The comparison shows that distinct peaks around 6.1 eV which are present in the tetramantane spectra are lacking in the pentamantane spectrum. The addition of a cage to either of the tetramantane structures to form [12(1)3] pentamantane thus results in an attenuation of large parts of the characteristic peaks. A possible reason for this observation is the lowering of the diamondoid's symmetry to the point group C_1 which results in the cancelation of all orbital degeneracy. A strong spectral resemblance can be observed for the three spectra in panel (a) in the region between 6.5 and 7.0 eV. Here, [123] tetramantane possesses a double peak that fades into a broad resonance with a shoulder in [12(1)3] pentamantane and finally into a single broad resonance in [121] tetramantane. This is the kind of behavior one would intuitively expect due to the fact that [12(1)3] pentamantane is in many regards a structural mixture of the two tetramantane and, in fact, contains both of them.

In panel (b) [1213] pentamantane is shown together with the spectra of [121] and [123] tetramantane. The absorption onset of [1213] pentamantane exhibits a small feature at 5.8 eV and a pronounced peak at 6.1 eV following the spectrum of [123] tetramantane remarkably well. The small feature at 5.8 eV for the L-shaped [1213] pentamantane is in agreement with the tentative ascription of this feature to the existence of a bend in the diamondoid structure. Another small feature at 6.3 eV is also found in the spectrum of [121] tetramantane. Beyond this point it is difficult to match spectral features of the pentamantane to those of a tetramantane. Interestingly, this behavior starkly contrasts the observations made for [12(1)3] pentamantane where similarities are scarce at the absorption onset and more abundant at higher energies. To overlay the spectra the spectra of the tetramantanes had to be offset a little further than in the above case of [12(1)3] pentamantane. Also it interesting to note that the shift among the tetramantanes varies slightly for both panels of Fig. 5.5 which indicates that electronic states responsible for the absorption onset shift independently from those at higher energies. This shows that different electronic levels, in general, exhibit a noticeably different response to changes in size and shape of the diamondoid.

The spectral analysis of all measured diamondoids shows that the optical response of diamond nanocrystals is intimately linked to their shape. In fact, the spectra of diamondoids of the same shape typically exhibit stronger similarities than for diamondoids of the same size. The most drastic effects are observed for the 1D, 2D, and 3D basic shapes. But the analysis of the additional pentamantane spectra showed that some of the observed spectral features can also be linked to certain geometric elements of a diamondoid structure.

The above results are the first to reveal a dominant influence of the shape of a nanocrystal over its size. While diamondoids are a unique system allowing such detailed experimental investigations no such limits apply to theoretical investigations. Yet, a noticeable influence of the shape on the optical properties of similarly sized nanocrystals has not been reported so far.

The results show that shape becomes a non-negligible design parameter when semiconductor nanostructures approach the sub-nanometer regime. As is seen by the huge differences in the behavior at the absorption onset the shape of the nanocrystal starts to govern the transition probability for the lowest-energy transitions. In a size regime of only a few nanometers, controlling the nanocrystal size will most likely no longer be sufficient to define the optical response of semiconductor nanocrystals. Along with greater challenges in materials design come great prospects that the control of the particle shape bears for the design of still more efficient nanophotonic devices. The direct interrelation of the shape and the optical response could thus give rise to a novel field of *shape-dependent photophysics*.

5.1. OPTICAL ABSORPTION

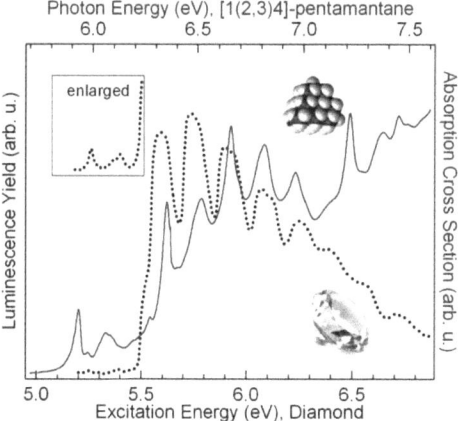

Figure 5.6: Comparison of the optical absorption of the 3D *basic shape* ([1(2,3)4]-pentamantane, solid line) with the absorption of high purity type IIa diamond (dotted line) [91]. The spectra are plotted over different axes which are shifted by 0.73 eV. The spectral inset shows the enlarged onset of the bulk diamond spectra.

5.1.2 *Shape* Not *Size* Defines The Smallest Diamond

In the above discussion it became obvious that the 3D basic shape, [1(2,3)4] pentamantane, is special in many regards. Not only does it possess by far the most strongly featured spectrum of all higher diamondoids it also, from a structural point of view, represents the smallest possible 3-dimensional nanodiamond.[3] Besides differentiating it from the 2D and 1D structures which have a fundamentally different geometric structure, the 3D basic shape earns its name from another perspective which is the comparison to bulk diamond.

In Fig. 5.6 the absorption of [1(2,3)4] pentamantane is compared to that of high purity type IIa diamond. The two spectra in Fig. 5.6 are plotted over different axes that are shifted by 0.73 eV to account for quantum confinement effects in the nanodiamond. When taking into account this size dependent shift the nanodiamond spectrum reproduces all characteristics of the bulk diamond spectrum with a striking precision. The five main resonances between 5.5 and 6.4 eV in

---

[3]Of course it could also be labeled 0-dimensional, as is often said of quantum dots (which are much larger). Here, *3-dimensional* refers to the fact that it could be uniformly extended to a 3-dimensional nanocrystal according to the growth scheme in Fig. 5.2.

the diamond spectrum all have their counterparts in the pentamantane spectrum and even the shoulder at 5.5 eV is reproduced by the nanodiamond. The strong intensity deviation at higher energies is due to the use of different measurement techniques. The nanodiamond absorption has been measured recording the transmission in the gas phase while the absorption of bulk diamond has been determined using luminescence excitation spectroscopy [91]. The decrease in the luminescence signal can be explained by a decreasing luminescence quantum yield at higher excitation energies. A similar effect is also observed for nanodiamonds, as discussed in section 5.2.

Additionally to the five main resonances, the spectrum of [1(2,3)4] pentamantane exhibits two strong features below the absorption edge of bulk diamond. The first of these features, which also defines the optical gap, can be attributed to a transition from the HOMO into the LUMO [90]. As will be seen in the following section, this transition is dipole-allowed. The LUMO is defined by the hydrogen surface passivation and lies lower in energy than the bulk states [24] as seen in section 2.3. The fact that the LUMO is due to the hydrogen surface of the diamondoid explains why in Fig. 5.6 the optical gap of the nanodiamond does not coincide with the bulk gap. In the bulk limit the fraction of the surface atoms becomes negligible and only states which belong to the internal crystal structure of the diamond are considered. However, there is also a spectral pre-edge structure in the bulk diamond spectrum which is enlarged in the spectral inset in Fig. 5.6. These features have previously been attributed to so-called N9-centers [89]. Interestingly, shape and spacing of the features is remarkably similar to the transitions in the nanodiamond spectrum which are attributed to the hydrogen surface states. A comparison of the surface-to-bulk ratio of the nanodiamond and the 1 mm diamond sample [91] showed that the intensity of these pre-edge features correlates with the surface-to-bulk ratio of the corresponding sample [79]. Therefore, surface states that may be accountable for the pre-edge features in diamond and optical transitions into the surface states should in principle be detectable in the bulk material.

It is important to point out that the spectral resemblance of bulk and nanodiamond go hand in hand with many structural similarities. Thus, even though unexpected at this minuscule size, the convergence of the optical properties found above does not come about arbitrarily. For example, [1(2,3)4] pentamantane exhibits T_d symmetry which is the largest finite subgroup of the bulk diamond's infinite O_h point group. The particular importance of symmetry for the optical

5.1. OPTICAL ABSORPTION

properties has been pointed out in section 3.3. The implications will be discussed in detail in section 5.1.3. Further, [1(2,3)4] pentamantane possesses four (111) facets which is one of the most prominent crystal surfaces encountered in the bulk material. These facets make up the entire surface of the nanocrystal. It is not clear, however, if the investigated specimen in Ref. [91] is cleaved in the (111) direction. And finally it is worthwhile pointing out that the observed spectral resemblance to bulk diamond exists exclusively for the 3D basic shape. No other investigated diamondoid bears a similar resemblance and 1D or 2D basic shapes or the two pentamantane structures discussed in section 5.1.1 exhibit none of the spectral features found in the bulk diamond spectrum even though they possess the same number of carbon atoms and the same underlying diamond structure. This corroborates the claim that only due to its shape [1(2,3)4] pentamantane can be considered a nanoscale model for diamond [52] and it underlines the eminent importance of a nanocrystal's shape for its optical properties.

5.1.3 Evolution of the Optical Gap

In principle, quantum confinement effects affect all electronic levels of a system. The focus of investigations, no matter if of experimental or theoretical nature, is typically laid on the size-dependence of the energy gap, which serves as a measure of the size-dependent effects. There are, however, various different definitions of the energy gap and the use of the term *energy* or *band gap* is, therefore, ambiguous. It very generally describes a concept of an electronic system where regions of occupied and unoccupied electronic states (*bands*) are separated by a region where no electronic states are present. This concept, which has been introduced in section 3.1, is more of a qualitative nature. It does not differentiate between various methods of determining the band gap.

In this work, optical absorption spectroscopy is used to determine size- and shape-dependent effects in the electronic structure. The energy gap determined from these data is thus, more specifically, the *optical gap* of the investigated diamondoids. For a quantitative discussion of the quantum confinement effects consistent and unambiguous criteria for determination of the optical gaps from the experimental data are needed. Further, a thoughtful definition will enable a direct comparison of the present data to theoretical results.

The optical gap is defined by the lowest energy at which photon-induced transitions between the occupied and the unoccupied states of a system occur. Theoretically this value is fairly easy to define: It is the lowest energy at which photon absorption occurs or, in a dipole approximation, the lowest energy at which a non-zero dipole-transition matrix element occurs.

Experimentally, however, the optical gap has to be determined as the lowest energy at which *a notable absorption signal* occurs. The two definitions are obviously not equivalent because the transition probability of single transitions could be arbitrarily small and virtually undetectable by experimental means and yet not be zero. Defining the experimental optical gap as the first major optical transition, i.e., the first clearly distinguishable peak in the spectrum would mean to neglect optical transitions with transition probabilities too small to be detectable. This approach does also not satisfyingly deal with the problem because a number of dipole transitions with vanishing transition probabilities could give rise to a steady increase in absorption and thus to an experimentally detectable optical gap. Also, the experimental optical gap should be comparable to a theoretically determined value.

As seen in section 5.1.1, the absorption spectra in Fig. 5.3 vary considerably in their appearance and, most notably, at the absorption onset. This makes it difficult to determine, compare and discuss the optical gaps of different diamondoid structures directly from the spectra. As a solution to the problem, the following approach is chosen: The optical gap is defined as the point where the integrated oscillator strength reaches a fraction x of its total value [92]. This definition of the optical gap as a fraction of the integrated oscillator strength has the advantage that it can also be applied to theory [93].

In order to determine the optical gap according to this method, the optical absorption is integrated over the entire spectral range. The resulting total integrated oscillator strength for each diamondoid structure is normalized to its value at the ionization potential [25]. The normalization to the ionization potential ensures that the determined value is not dependent on the number of atoms of the system and thereby inherently size-dependent. The energy at which the normalized value reaches a predefined fraction x, which has yet to be determined, is taken as the optical gap. This way a standardized method to determine the optical gap from experimental data is established.

The integrated absorption signals for the diamondoids which have been assigned to one of the categories in section 5.1.1 are shown in Fig. 5.7. The background has

5.1. OPTICAL ABSORPTION

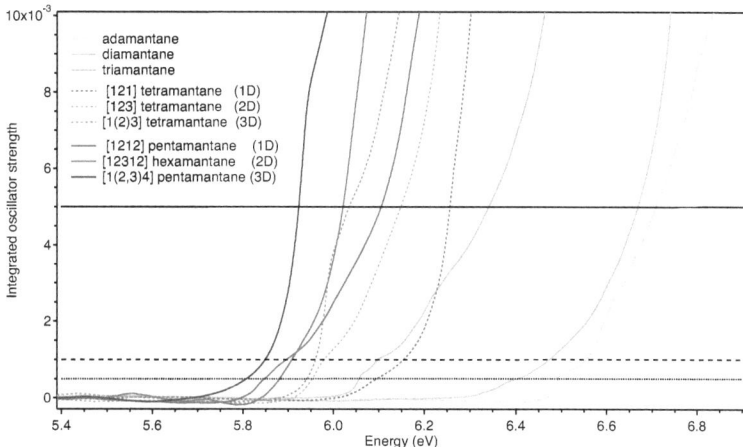

Figure 5.7: The integrated absorption spectra of the diamondoids categorized shown in Fig. 5.2. The curves are normalized to their value at the corresponding ionization potential. Different threshold values for the optical gap are indicated by a dotted ($x = 5 \cdot 10^{-4}$), a dashed ($x = 1 \cdot 10^{-3}$), and a solid line ($x = 5 \cdot 10^{-3}$).

been subtracted prior to integration so that the baselines of the optical spectra oscillate evenly around the x-axis. This is necessary for a determination of the optical gap with high precision because any offset in the original spectrum will accumulate in the integration of the spectrum and lead to a non-constant background contribution. A minimal offset in the positive y-direction of the optical spectrum, as it may occur, e.g., from slightly offset transmission scans for the full and the empty cell, will result in a steady increase in the integrated signal and possibly trigger the threshold for the optical gap before any intrinsic absorption occurs. Also in some cases a slightly tilted baseline needs to be assumed, e.g., in the case of adamantane in Fig. 5.8, to avoid the adverse influence of experimental effects, such as the fading of synchrotron beam intensity over time. For all diamondoids the integrated absorption signal in Fig. 5.7 abruptly leaves the baseline which distinguishes it from the small scale noise and unambiguously indicates the absorption onset.

The choice of x is necessarily arbitrary in a sense that no distinguished value for x exists. In a theoretically pure spectrum the gap could be defined as the point where the signal leaves the baseline. In the experimental case this is not

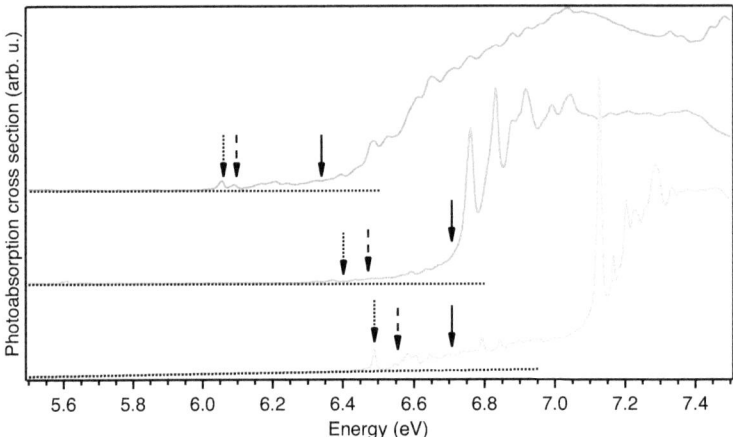

Figure 5.8: The optical spectra of adamantane (bottom), diamantane (middle), and triamantane (top) are shown. The different style arrows indicate the optical gaps for the different fractions of the integrated oscillator strength as shown in Fig. 5.7: dotted ($x = 5 \cdot 10^{-4}$), dashed ($x = 1 \cdot 10^{-3}$), and solid ($x = 5 \cdot 10^{-3}$). Only the smallest of those three values is able to 'detect' the first sharp transition in adamantane and triamantane. For diamantane the situation is less obvious.

possible, as mentioned above, because of signal noise and other adverse effects. Thus, a finite threshold x has to be set. Three different choices for the threshold defining the optical gap - $x = 5 \cdot 10^{-4}$, $x = 1 \cdot 10^{-3}$, $x = 5 \cdot 10^{-3}$ - are indicated by a dotted, a dashed, and a solid line respectively. The optical gaps which result from the three different choices for the threshold value are listed for all diamondoids in Tab. 5.1. All three threshold lines shown in Fig. 5.7 are clearly above the noise level for the diamondoids. $x = 5 \cdot 10^{-4}$ is the smallest possible value that is still reliably distinguishable from the background noise level. A further reduction of the threshold would lead to ambiguities as an analysis of possible errors demonstrates.

To illustrate possible sources of error the example of the lowest data quality, [12(1)3] pentamantane, is used. The same error assessment was used for all of the diamondoids and typically yielded considerably lower errors.

Prior to integration the baseline of the spectrum has to be chosen in order to get rid of background contributions. Depending on the choice of this baseline the results for the integrated optical spectrum vary greatly, giving rise to an

5.1. OPTICAL ABSORPTION

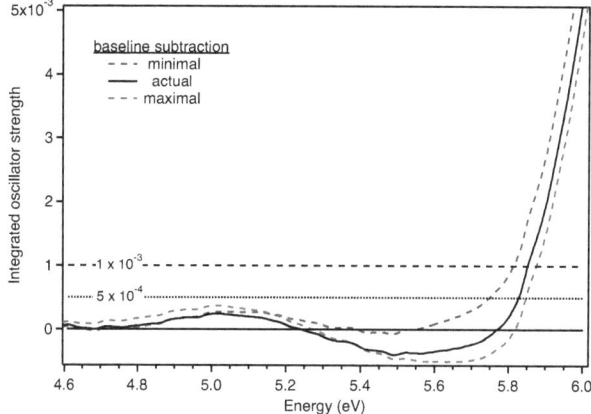

Figure 5.9: The determination of the optical gap for [12(1)3] pentamantane. This graph illustrates the uncertainty of the method due to the choice of the baseline for the integration of the spectrum. The results for two extreme choices are shown together with the actual choice. Variance between the two extreme values for the optical gap for [12(1)3] pentamantane is 100 meV for the $x = 5 \cdot 10^{-4}$ threshold and smaller for larger thresholds. It is much smaller for most of the other diamondoids. The example of [12(1)3] pentamantane which has the poorest data quality has been chosen to demonstrate the principle.

uncertainty in the value for the optical gap. The effect of the choice of different baselines is visualized in Fig. 5.9 using the example of [12(1)3] pentamantane. The integrated and normalized oscillator strength for the actual choice of the baseline (solid line) is contrasted with two extreme choices which result in minimal (dashed line, mostly above) and maximal (dashed line, mostly below) values for the optical gap. Indicated by a dotted and a dashed horizontal line are threshold values of $x = 5 \cdot 10^{-4}$ and $x = 1 \cdot 10^{-3}$, respectively. These thresholds result in values of 5.828 eV (5.748/5.848) and 5.852 eV (5.814/5.876)[4] for the actual (minimal/maximal) optical gaps, respectively. The variance between minimal and maximal value for the optical gap is exactly 100 meV for a threshold of $x = 5 \cdot 10^{-4}$. For $x = 1 \cdot 10^{-3}$ ($x = 5 \cdot 10^{-3}$) it is 62 meV (38 meV). The maximum deviation from the actual optical gap of [12(1)3] pentamantane of 5.828 eV is thus 80 meV.

---

[4]These values are the direct readout from the graphs. Of course, they are not meant to be accurate to four significant digits.

A second possible source of error is the propagation of errors made in the determination of the ionization potential. As mentioned above, the integrated spectra are normalized to their value at the corresponding ionization potential. This source of error becomes more significant, however, for diamondoids where the ionization potentials has not yet been determined. Besides [1213] and [12(1)3] pentamantane which were measured only very recently, this applies also to [1212] pentamantane. For those structures the ionization potential has been assumed to be 8.07 eV, i.e., the same as for [1(2,3)4] pentamantane [25]. Normalizing the integrated signal to ionization potentials that lie 0.15 eV higher or lower than the assumed value results in a shift of the optical gap of less than 0.01 eV. Thus, even for diamondoids with unknown ionization potentials the propagating error in the determination of the optical gap remains negligible. The ionization potentials for most of the investigated diamondoids have been determined to a high precision [25, 37] and the uncertainty in the ionization potentials results in additional uncertainties of only ~1 meV in the optical gaps.

The error in the optical gap of [12(1)3] pentamantane is estimated to be smaller than the sum of the two main errors, i.e., 0.08 eV from varying the baseline and 0.01 eV from the uncertainty in the ionization potential. The resulting comparably large error of 0.09 eV is due to the lower data quality for [12(1)3] pentamantane and only applies to [1213] and [1212] pentamantane. The error for the optical gaps of all other diamondoids has been determined the same way and is no larger than 0.03 eV.

To demonstrate how the optical gaps which result from the different thresholds compare to the spectra, the spectra of adamantane through triamantane are shown in Fig. 5.8. The arrows indicate the optical gaps which result from the different choices for the threshold values: arrows with dotted, dashed and solid arrow lines stand for fractions of $x = 5 \cdot 10^{-4}$, $x = 1 \cdot 10^{-3}$, and $x = 5 \cdot 10^{-3}$ of the total oscillator strength, respectively. The baseline for each spectrum is given by a dotted line. In the case of adamantane only the smallest of the three threshold is able to 'detect' the first optical transition, i.e., to correctly determine the optical gap. For diamantane the situation is less obvious because no intense optical transition exists at the absorption onset. Yet the spectrum clearly leaves the baseline before the appearance of first pronounced peaks around 6.7 eV. For triamantane on the other hand a distinct first transition exists. It coincides with the optical gap determined from the smallest of the suggested threshold values. Only the smallest of the three thresholds is thus able to 'detect' the lowest energy

5.1. OPTICAL ABSORPTION

diamondoid name	optical gap (eV) @ threshold x		
	@ $5 \cdot 10^{-4}$	@ $1 \cdot 10^{-3}$	@ $5 \cdot 10^{-3}$
adamantane	**6.492**	6.555	6.707
diamantane	**6.404**	6.473	6.668
triamantane	**6.057**	6.098	6.340
[121] tetramantane	**6.097**	6.154	6.255
[123] tetramantane	**5.953**	5.979	6.147
[1(2)3] tetramantane	**5.941**	5.960	6.034
[1212] pentamantane	**5.847**	5.896	6.104
[12312] hexamantane	**5.881**	5.907	6.020
[1(2,3)4] pentamantane	**5.807**	5.849	5.923
[1213] pentamantane	**5.787**	5.837	5.985
[12(1)3] pentamantane	**5.828**	5.852	5.995

Table 5.1: Experimentally determined optical gaps of diamondoids for different threshold values. The smallest value, $5 \cdot 10^{-4}$, is the best choice for the experimental optical gap as explained in the text.

transitions. In the obvious cases, these coincide with an intuitive visual readout of the gaps from the spectra. This shows that an extremely high data quality and precision in the analysis is required to derive meaningful optical gaps using the above method. In the cases of triamantane and, more drastically, adamantane it became obvious that the larger of the threshold values indicated in Fig. 5.7 fall short of correctly determining the optical gap. In the case of diamantane an intuitive readout of the optical gap from the experimental data is highly ambiguous. The curves of the integrated oscillator strength in Fig. 5.7 show that the absorption of diamantane sets in much more slowly than those of other diamondoids which makes the value for diamantane more error-prone. However, Fig. 5.7 also shows that the absorption signal clearly leaves the baseline and that the optical gap has to be placed at least in the vicinity of the derived value. The derived energy for the optical gap, therefore, provides a value that is consistent with the values for all other measured diamondoids enabling a meaningful discussion of the optical gap.

With the procedure above, a convincing and consistent method to determine the optical gap has been introduced. The quality of the optical data allows to set the threshold for the optical gap at a fraction as low as $x = 5 \cdot 10^{-4}$ of the normalized oscillator strength. This threshold value, which is small enough to detect

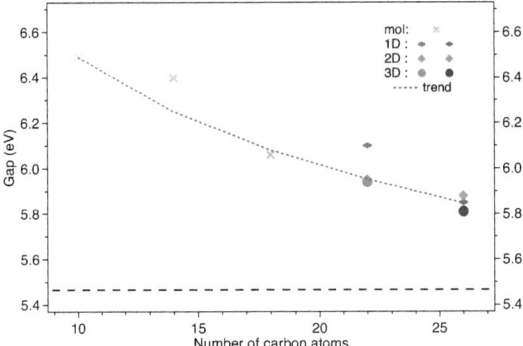

Figure 5.10: Optical gaps of of diamondoids as a function of cluster size. The dashed line indicates the bulk diamond band gap. An overall trend of a decreasing optical gap with increasing diamondoid size is observed. Deviations from the trend appear for diamantane (14 carbon atoms) and [121] tetramantane (1D, 22 carbon atoms).

noticeable but weak transitions within the spectra, is comparable to values used in theoretical studies [93].

The optical gaps are shown as a function of cluster size in Fig. 5.10. The optical gaps exhibit an overall trend of decreasing optical gaps with increasing cluster size as is expected from the quantum confinement model. The optical gap values of diamantane ($C_{14}H_{20}$) and [121]tetramantane (1D transitional structure, 22 carbon atoms) exhibit a clear deviation from a simple scaling of the gap with the number of atoms. These irregularities coincide with the lack of pronounced spectral features at the absorption onset observed in Fig. 5.1.

To investigate the correlation between the gap value and the behavior of the spectrum at the absorption onset, dipole selection rules governing optical transitions are considered. As explained in section 3.3, one can identify dipole-forbidden transitions by pure symmetry considerations. Note that the application of group theoretical concepts to experimental results requires the physical system of interest to be perfectly defined in its structure and the sample to be absolutely pure. In the case of diamondoids, where the prerequisites are given, group theory provides a very useful tool to further investigate the optical gaps.

As a first step, the symmetry properties of the molecular orbitals are calculated

5.1. OPTICAL ABSORPTION

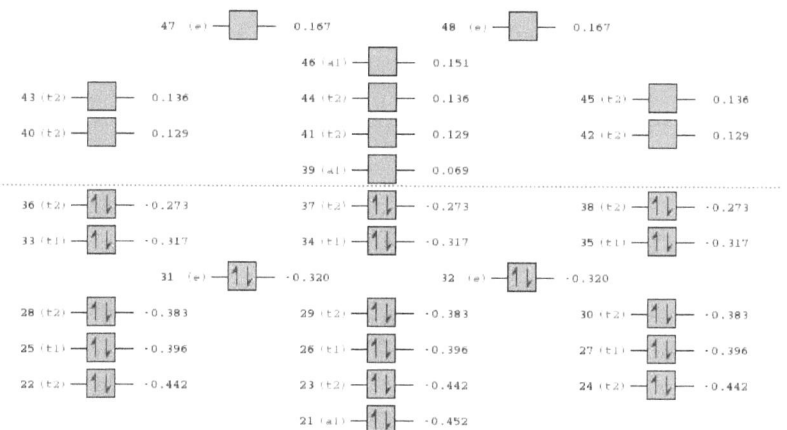

Figure 5.11: The highest occupied and lowest unoccupied molecular orbitals of adamantane as computed using GAUSSIAN03 and displayed by GaussView. Degenerate orbitals are shown next to each other. The binding energy (in hartree) and the symmetry of each orbital are indicated.

for all investigated diamondoid structures. Subsequently, the group theoretical concepts which have been introduced in section 3.3 are applied to diamondoids to identify dipole-forbidden transitions.

Quantum chemical methods, introduced in section 3.2, are employed to compute the electronic structure of the investigated diamondoids. The symmetries of the molecular orbitals are computed at 6-31G* level of theory using the B3LYP hybrid functional (compare chapter 3.2) as implemented in GAUSSIAN03 [70]. In Fig. 5.11 a scheme of the calculated molecular orbitals for adamantane is shown as an example. The computed energy (in hartree[5]) is shown to the right of each orbital. The corresponding irreducible representation (compare section 3.3) is noted on the left hand side. Degenerate orbitals are displayed next to each other. To verify the reliability of the method the orbitals of adamantane have also been computed using more complex basis sets (B3LYP/6-311++G**, B3LYP/aug-cc-pVQZ and PBE/aug-cc-pVQZ). All of the methods yielded the same results for the type of the orbital symmetry and for their energetic order. Considerable differences exist for the absolute binding energies of molecular orbitals but this does not affect their order. The calculations for adamantane show that, despite the fact that it is clearly insufficient to quantify the orbital energies, the rather basic

---

[5] 1 hartree = 27.21 eV

CHAPTER 5. RESULTS & DISCUSSION - PART I: PRISTINE DIAMONDOIDS

level of theory of B3LYP/6-31G* which is employed here is sufficient to derive a qualitative understanding of the energy levels. This qualitative understanding of the molecular orbitals, namely, the knowledge of their irreducible representations allows for the application of the fundamental selection rules to diamondoids.

In the following the irreducible representations of the molecular orbitals are compared with the dipole-selection rules for each group. The irreducible representations of the three highest occupied and lowest unoccupied orbitals are listed in table 5.2. Dipole-forbidden transitions for each of the relevant molecular point groups are identified according to the example given in section 3.3 and are listed in the Appendix.

Adamantane and [1(2,3)4] pentamantane belong to the point group T_d. For this particular point group with the irreducible representations A_1, A_2, E, T_1, and T_2 transitions between A_1, A_2, and E are forbidden in every combination ($A_1, A_2, E \not\leftrightarrow A_1, A_2, E$). Further, $A_1 \not\leftrightarrow T_1$ and $A_2 \not\leftrightarrow T_2$. For adamantane comparison with the orbital representations in table 5.2 shows that transitions from HOMO and HOMO-1 (both T_2) into the first three unoccupied states are allowed. Among the listed levels only the transition from the HOMO-2 to the LUMO ($E \rightarrow A_1$) is forbidden. For [1(2,3)4] pentamantane it is also only the transition HOMO-2 (E) $\rightarrow$ LUMO (A_1) which is forbidden. Thus, there are no relevant effects of the dipole selection rules on the optical gap of either adamantane or [1(2,3)4] pentamantane. This is in agreement with the spectra of adamantane or [1(2,3)4] pentamantane in Fig. 5.3 which both spectra feature a sharp peak at the absorption onset.

Diamantane and [12312] hexamantane exhibit D_{3d} symmetry. For this particular point group all representations have a defined parity which is either even ('g' for *gerade*) or odd ('u' for *ungerade*). In the list of the forbidden transitions in the Appendix the letters g and u stand short for the set of even and odd representations, respectively. Trivially, dipole transitions between molecular orbitals of the same parity are forbidden. Further, $A_{1g} \not\leftrightarrow A_{1u}$ and $A_{2g} \not\leftrightarrow A_{2u}$. It can be deduced in combination with table 5.2 that for diamantane several transitions between molecular orbitals near the band edges are dipole-forbidden. Among them, most notably, HOMO→LUMO and HOMO-1→LUMO. This can explain the deviation of diamantane from the trend of the optical gaps which is observed in Fig. 5.10. It also explains the lack of sharp spectral features at the absorption onset of diamantane.

5.1. OPTICAL ABSORPTION

For [12312] hexamantane the HOMO-LUMO transition is forbidden as well. In fact, transitions from all of the three highest occupied orbitals into the LUMO are dipole-forbidden. Nevertheless, a notable deviation from the trend for the optical gaps is not observed. However, a look at Fig. 5.3 reveals that the spectrum of [12312] hexamantane (2D basic shape) does not possess a peak at the absorption onset. Such a peak is observed, e.g., for the 3D basic shape. But also for the 2D transitional structure a small peak that is present at the absorption onset which vanishes completely in [12312] hexamantane.

Triamantane and [1212] pentamantane belong to the point group C_{2v}. In consequence for their transitions holds $A_1 \nleftrightarrow A_2$ and $B_1 \nleftrightarrow B_2$. This means that for both triamantane and [1212] pentamantane, among the electronic levels which are close to the band edges, only the transitions HOMO $\rightarrow$ LUMO+2 and HOMO-2 $\rightarrow$ LUMO+1 are forbidden. The HOMO-LUMO transitions are not affected for either of the diamondoid structures exhibiting C_{2v} symmetry. In the triamantane spectrum this manifests itself in the presence of a peak defining the optical gap. For [1212] pentamantane things are less clear due to the lower data quality which prohibits a definite identification of small and yet pronounced peaks. A peak at the absorption onset could be present, however, the data quality does not allow a definite judgement.

[121] tetramantane is the only investigated diamondoid with C_{2h} symmetry. This point group possesses irreducible representations of defined parity and thus a change of parity is required by the dipole selection rules. No extra restrictions apply. A look at table 5.2 reveals that transitions from all three highest occupied levels into the LUMO are dipole forbidden. First transitions appear from the HOMO and the two subsequent occupied orbitals into the LUMO+1 and the LUMO+2. Again, just like for diamantane, the deviation of [121] tetramantane from the trend of the optical gaps in Fig. 5.10 can thus be explained by symmetry considerations.

[123] tetramantane has one of the lowest symmetries among the investigated diamondoids: It is a member of the point group C_2. For [123] tetramantane - and for all other quantum mechanical systems with C_2 symmetry - transitions between all existent levels are dipole-allowed. In the absorption of [123] tetramantane a small but noticeable absorption peak is observed at the onset of the spectrum indicating a notable HOMO-LUMO transition. Within the 2D structural family this feature vanishes for the basic shape, [12312] hexamantane, for which the HOMO-LUMO

transition is forbidden due to symmetry costraints. The difference here lies in the group theoretical restrictions, as seen above.

[1(2)3] tetramantane possesses the highest symmetry among the transitional structures, namely C_{3v} symmetry[6]. For structures of C_{3v} symmetry only transitions between orbitals with the representations A_1 and A_2 are forbidden: $A_1 \not\leftrightarrow A_2$. This group theoretical restriction does not, however, affect any of the lowest energy transitions as the orbital symmetries listed in table 5.2 transform as A_2. This fits well to the observation of a very pronounced double peak defining the optical gap. In general, the 3D geometric family is by far the most symmetric of the three geometries. Besides the possibility of dipole-forbidden transitions the high symmetry gives rise to a high degree of degeneracy of the electronic levels as can be seen in table 5.2: several doubly degenerate levels (E) exist for [1(2)3] tetramantane and a large number of triply degenerate levels ($T_{1/2}$) for [1(2,3)4] pentamantane. This explains the large amount of spectral structure in the 3D branch of Fig. 5.3.

Finally, [12(1)3] pentamantane and [1213] pentamantane belong to the molecular point group C_1 which does not contain any symmetry elements (except for the identity relation). This means none of the transitions between electronic levels are restricted by symmetry. Therefore, these diamondoids are not listed in table 5.2.

In the last column of table 5.2 the findings for the HOMO-LUMO transition for each diamondoid are summarized. A check mark identifies an allowed HOMO-LUMO transition and a cross a dipole-forbidden one. The deviations from the trend for the optical gap which have been observed for diamantane and [121] tetramantane coincide with dipole forbidden-transitions between the HOMO and the LUMO for these structures. The lowest transitions for the 2D basic shape, [12312] hexamantane, are also symmetry-forbidden. Similar to diamantane and [121] tetramantane, this is also reflected in the spectrum which also exhibits a smooth and featureless absorption onset. The absolute gap value, however, exhibits only a minor deviation from the other two basic shapes. The fact that of all samples the 2D basic shape is measured at the highest temperature provides a possible explanation: Phonon-assisted transitions may lead to a lowering of the measured gap. Also geometric distortions due to thermal activation could lift the strict dipole-selection rules that are in place for the other two structures and lead

---

[6]Note that for this group the detailed deduction of the dipole-forbidden transitions is demonstrated as an example in section 3.3.

5.1. Optical Absorption

Diamondoid name	Point group	Occupied levels			Unoccupied levels			HOMO-LUMO
		-2	-1	HOMO	LUMO	+1	+2	
adamantane	T_d	E	T_2	T_2	A_1	T_2	T_2	✓
diamantane	D_{3d}	E_u	E_g	A_{1g}	A_{1g}	A_{2u}	E_u	✗
triamantane	C_{2v}	B_1	A_1	B_2	A_1	B_2	B_1	✓
[121] tetramantane	C_{2h}	B_g	A_g	A_g	A_g	B_u	A_u	✗
[123] tetramantane	C_2	A	A	B	A	A	B	✓
[1(2)3] tetramantane	C_{3v}	E	A_1	E	A_1	E	A_1	✓
[1212] pentamantane	C_{2v}	B_1	A_1	B_2	A_1	B_2	B_1	✓
[12312] hexamantane	D_{3d}	A_{1u}	A_{1g}	E_g	A_{1g}	A_{2u}	E_u	✗
[1(2,3)4] pentamant.	T_d	E	T_1	T_2	A_1	T_2	T_2	✓

Table 5.2: This table lists the molecular point group of each diamondoid together with the irreducible representations of the three highest occupied and the three lowest unoccupied electronic levels. The orbital representations are derived from computations at 6-31G* level of theory using GAUSSIAN03. In the last column it is indicated whether or not the HOMO-LUMO transition is dipole-allowed.

to a weak coupling of the HOMO and the LUMO.

The allowed HOMO-LUMO transitions for all other diamondoid structures coincide with an accentuated single or double peak at their absorption onset defining the optical gap. This spectral feature can thus be assigned to transitions from the highest occupied states into the LUMO which is defined by the hydrogen surface [24]. This interpretation is in agreement with recent calculations predicting optical gaps of larger hydrogen-passivated nanodiamonds to lie below the bulk value due to such surface states [94]. Further, it is worthwhile mentioning that the above observations are a recognition of structural properties of diamondoids in their optical spectra. The correlation of geometry and optical response provides an additional and direct proof of the quality of the samples.

Next, the experimental data for the optical gaps are compared to data from different studies. In Fig. 5.12 different experimental results which yield information about the energy gap in diamondoids are summarized. In panel (a) the optical gaps of diamondoids are compared to gap values which have been derived from the combination of band edge data for the occupied and the unoccupied states [36]. Willey et al. used soft x-ray emission (SXE) measurements to determine the

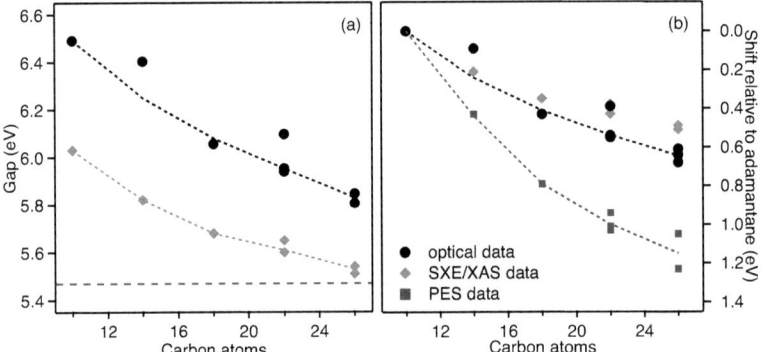

Figure 5.12: (a) Comparison of different experimental values for the energy gap of diamondoids. The optical gaps determined in this work are compared to values for the energy gap derived from a combination of SXE and XAS [36]. (b) Comparison of size-dependent shift relative to adamantane. PES data [37] are also included because the LUMO is constant in energy.

relative position of the occupied states and combined the data with x-ray absorption spectroscopy (XAS) of the unoccupied states [24]. Both techniques probe the corresponding states with respect to the C1s core-levels which thus provide a common reference to derive values for the energy gap. To estimate values for the energy gap of diamondoids, the XAS and SXE spectra of bulk diamond were acquired and calibrated to the well-known gap of bulk diamond at 5.47 eV. The resulting gap values, which provided a first experimental proof of quantum confinement effects in diamondoids, are considerably smaller than the optical gaps. The SXE/XAS method turns out gap values that are constantly about 0.5 eV lower than the optical gaps. The data suffer from an inherently low resolution of only 0.5 eV and beam damage on the condensed samples occurred during the measurements which made a precise determination of the gaps difficult. In addition, the measurements were performed on condensed samples and the effect of the particle-particle interaction remains unclear. An exact agreement of the gaps derived from the SXE/XAS measurement with the optical gaps cannot be expected because both SXE and XAS in fact probe systems with core-hole in either the initial or in the final state, respectively. Therefore, it is difficult to conclude whether the observed deviations are due to the error bars of both methods or, which seems more likely, to the fact that they probe different states.

5.1. OPTICAL ABSORPTION

The trends for the two methods are in very good agreement, however, as can be seen from panel (b) which shows the size-dependent shift relative to adamantane. For diamondoid structures with a forbidden HOMO-LUMO transition the difference between the two methods is larger. This can be expected because the dipole-selection rules do not impose the same restrictions upon the conducted band edge measurements. Therefore, the lack of irregularities in the trend of the SXE/XAS measurements is additional evidence that the deviation from the trend in the optical gaps is due to symmetry and not to the inherent electronic structure of the diamondoids.

Also added to panel (b) is the size-dependent shift of the valence band edge derived from photoelectron spectroscopy (PES) [26]. The size dependence of the valence band edge provides another measure for the opening of the band gap because the conduction band is fixed in energy as seen in section 2.3. In principle, the size-dependence of the valence states should thus give the size-dependence of the energy gap. Both, the optical gaps and the energetic position of the conduction band edge have been measured with great accuracy [90, 26] and error bars on the experimental data cannot be held responsible for the obvious deviation. The differences in the size-dependence thus has to be attributed to effects of the corresponding final states. For the PES data a comparison to the core levels identified size-dependent screening effects as a possible source of distortion [37]. This effect, however, lessens the size-dependent shift in the valence band edge by 0.25 eV at most. This is only half of the discrepancy between the size effects in the valence band edge and the optical gap. Therefore, size-dependent effects in the optically excited state have to be present which counteract the quantum confinement effects on the band edges and, effectively, diminish the optical gap. The main difference of optical absorption to PES, where the final state of the diamondoid is ionic, is the creation of an electron-hole pair. The magnitude of the electron-hole interaction is known to be highly size dependent on the nanometer scale [95]. The electron-hole binding energies increase with decreasing cluster size as is observed for the difference between absorption and PES measurements. Thus, size-dependent electron-hole (or exciton) binding energies provide an explanation for the difference in the size dependence of valence band edge and optical gaps of diamondoids. Simultaneously, the observed deviation between the trends in PES and optical measurements provides a maximum value for the size-dependence of the electron-hole binding energy in diamondoids.[7]

---

[7]The SXE/XAS data are not discussed in this context because of the fact that condensed

The experimental values derived above can be directly compared to computational values if a similar theoretical approach is used. As mentioned before, different definitions of the energy gap exist. Most notably the optical gap must not be confused with commonly computed HOMO-LUMO gap. A brief disambiguation is given to clarify the systematic problems which a comparison of optical and HOMO-LUMO gap yields.

The HOMO-LUMO gap is defined as the energy difference of the highest occupied and the lowest unoccupied electronic state. It is, first and foremost, a theoretical concept which is typically applied to systems in their electronic ground state. In theory, the determination of the HOMO-LUMO gap is fairly simple and straight forward. Typically, the geometry of the investigated system is optimized using quantum chemical software, such as GAUSSIAN or TURBOMOLE (compare section 3.2). Then, the electronic structure of the particle in its optimized geometry is computed which, among other, turns out number values for the energy of all occupied and unoccupied electronic levels, i.e., the molecular orbitals. The quality of these predictions depends strongly on the employed method and the complexity of the basis set which is used. In an experiment single electronic levels are not directly accessible. To gain information about the electronic states of a system, such as the HOMO and the LUMO, the system has to be manipulated. Typically, an electronic excitation of one kind or another is used to measure the difference between two energy levels. Thus, in an experiment the energy of the HOMO and the LUMO will always be determined with respect to another electronic level. This level could, for the determination of the HOMO, be given by the conduction states (optical absorption), the core levels (soft x-ray emission), or the vacuum level (photoelectron spectroscopy). To compare the experimental HOMO energies for different systems one has to make sure that the level of reference does not shift. Another experimental difficulty arises from the fact that measuring the difference between to electronic levels always implies that the initial and the final state of the investigated system are not the same. In particular, this means that measuring the LUMO of a system in the ground state is impossible because either the initial or, typically, the final state will be an excited state, i.e, have an electron in the investigated electronic level. This, naturally, induces

---

samples were investigated and because of the large error bars which would only allow questionable conclusions. Note, however, that both SXE and XAS probe excitonic states as well (core exciton).

5.1. OPTICAL ABSORPTION

changes in the electronic structure of the system. The LUMO is therefore a so-called virtual level which does not exhibit physical reality in a strict sense. As a consequence it is, strictly speaking, not possible to experimentally determine the HOMO-LUMO gap.

The optical gap, on the other hand, is given by the lowest energy at which the absorption of a photon occurs. It therefore is defined by a transition process between the ground state and the first-excited state of the system. An experimental definition has been given at the beginning of this section. In theory, there are at least two different ways to derive a value for the optical gap. The first one is to separately compute the total energies of the ground and the excited state of a system. The minimum photon energy that is required for an optical transition is then given by the difference in total energy of the excited and the ground state.[8] This method, however, neglects the fact that a transition from the ground to the excited state is required. As explained in section 3.3 transitions between certain electronic levels can be dipole-forbidden in highly symmetric systems. Further transition matrix elements can simply be zero. Therefore, a second more sophisticated theoretical approach to the optical gap calculates the coupling between the electromagnetic field of the incoming light with the electronic levels and derives probabilities for photon-induced transitions between the different electronic levels. The latter, more complex theoretical approach includes all relevant effects and the results are without restrictions comparable to the experiment.

In Fig. 5.13 the experimental values are compared to theoretical values for optical gaps. The theoretical gap values in Fig. 5.13 have been computed using methods which are explicitly designed to yield optical gaps. The quantum Monte Carlo (QMC) calculations are from the year 2005 and predict values for only three diamondoid structures: adamantane, diamantane and [1(2,3)4]pentamantane [31]. All of the values by far overestimate the quantum size effects in diamondoids, for [1(2,3)4]pentamantane even by as much as 300%. Furthermore, the irregularity for diamantane is not observed in the QMC data because the optical gap has simply been calculated as the difference between the ground and the excited states. The transition probabilities have not been included in the calculation [31].

The other set of theoretical data used time-dependent density functional theory with the PBE0 hybrid functional (TDPBE0) to compute the optical gaps [96]. The computations were able to reproduce the experimental data with great pre-

---

[8]This is true for purely electronic transitions. Possible participation of phonons is neglected.

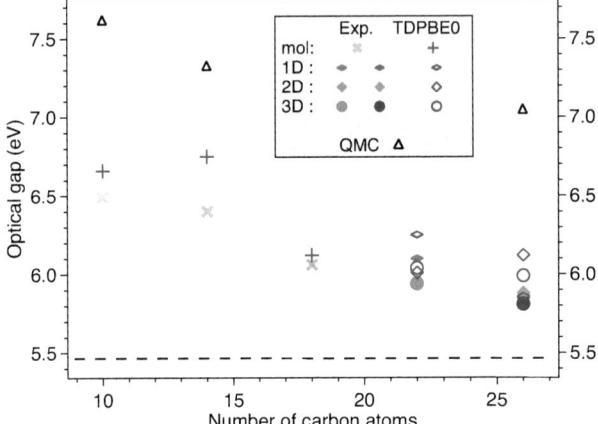

Figure 5.13: Optical gaps of diamondoids. Comparison of the experimental values with two sets of theoretical values which explicitly calculated optical gaps. The theoretical approaches used quantum Monte-Carlo methods (QMC) [31] and time-dependent density functional theory with the PBE0 hybrid functional (TDPBE0) [96]. For the experimental values and the TDBPE0 calculations the different structures are indicated by different symbols.

cision and the values for the gaps agree very well with the experimental data as seen in Fig. 5.13. Obviously the PBE0 hybrid functional provides a treatment that is more suitable for the description of the optical properties of diamondoids. It is of considerable value to hold a theoretical method that has proven that it is capable of reliably computing the optical properties of diamondoids and maybe of other, similar systems.

The two methods both come to the conclusion that the LUMO of diamondoids is delocalized at the surface. The more recent TDPBE0 investigation further identifies the LUMO as a 3s-like Rydberg state (compare). The radial distribution function of the LUMO, computed in Ref. [96], are shown in the insets in Fig. 5.14 (top). They demonstrate that the first optical transition in diamondoids is given by a 3s-like Rydberg excitation. These findings are different from those of earlier studies [24, 31] which attributed the LUMO a *surface nature*. However, in these studies the interpretation of the LUMO as a surface state was emphasized to be of strictly empirical nature. A solid theoretical understanding, which now has been reached by the means of combining experimental and theoretical findings,

5.1. OPTICAL ABSORPTION

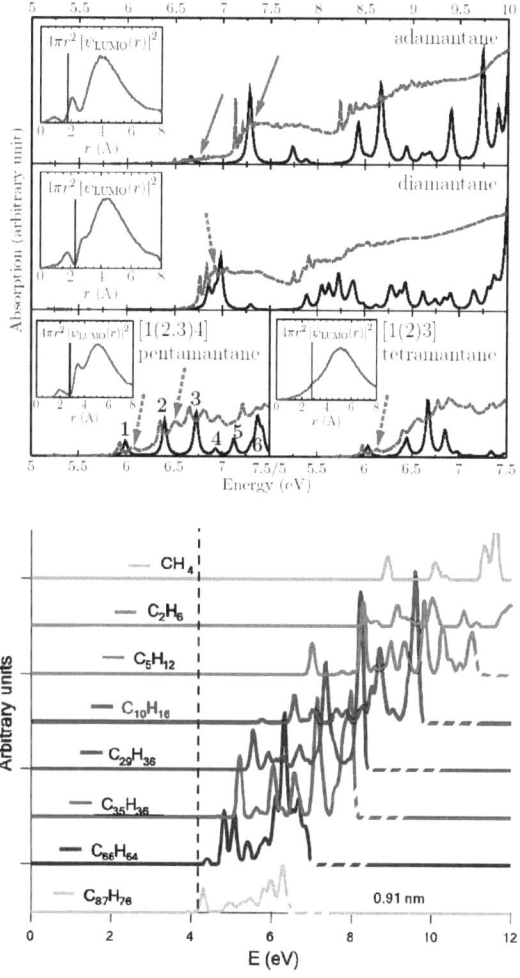

Figure 5.14: Top: Calculated optical spectra of diamondoids using time-dependent density functional theory with the PBE0 hybrid functional (TDPBE0) [96]. The computed spectra show a good agreement with the measured spectra and the reproduction of the features, especially for [1(2,3)4] pentamantane, is excellent. The insets show the radial distribution function of the LUMO and the arrows mark vibrational bands which are not reproduced by the electronic structure calculations.

Bottom: Spectra of different hydrogenated nanodiamonds as calculated with TDLDA [97]. Neither the spectrum of adamantane ($C_{10}H_{16}$) nor any of the other spherical nanodiamond structures bear recognizable resemblance with the experimental spectra.

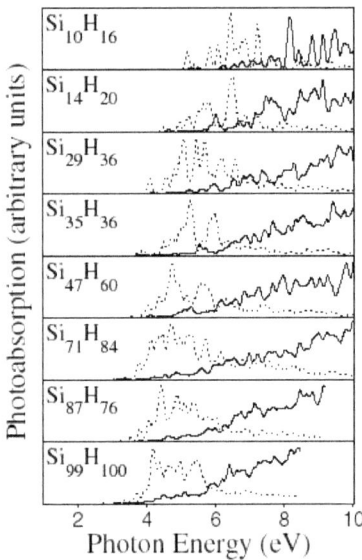

Figure 5.15: The optical spectra of hydrogenated silicon clusters [93] calculated using a time-dependent LDA (solid line). The time-independent approach (dotted line) has been added by the authors only for comparison and cannot be expected to yield accurate results. The first two structures are Si analogs of adamantane and diamantane, the rest are spherical nanocrystals. Drastic changes in the spectral structure, as observed in diamondoids, are not present in Si-nanocrystals.

has not been claimed. Exceptions to the 3s-like Rydberg nature of the lowest transitions are diamantane, [121] tetramantane and [12312] hexamantane, where these transitions are symmetry-forbidden. In these structures the first transition is identified to have 3p-like character.

In addition to calculating the energies of the optical gaps attempts have been made to compute the optical spectra of diamondoids. The recent TDPBE0 calculations were able to reproduce the optical spectra of diamondoids with great precision as shown in Fig. 5.14 (top). Especially for [1(2,3)4] pentamantane the agreement of the measured and the calculated spectrum is excellent. An earlier study used time-dependent density functional theory within the local density approximation (TDLDA) to calculate the optical spectra of small nanodiamonds [97]. These calculations fail to reproduce the well-known spectra of adamantane

5.1. OPTICAL ABSORPTION

[32] as is apparent in the bottom panel of Fig. 5.14. In the same work the quantum size effects compared to the bulk reference (dotted line in Fig. 5.14) are largely overestimated for adamantane and other structures similar to the ones in this work.

The closest related system to the presently investigated diamond clusters are hydrogen passivated silicon nanocrystals. Various theoretical approaches have been applied to the problem of optical gaps and spectra for these systems [98, 93, 99]. The optical spectra calculated for several hydrogenated Si-nanocrystals using TDLDA are shown in Fig. 5.15 (solid line) [93]. The time-independent approach (dotted line) has been added by the authors only for comparison and cannot be expected to yield accurate results. The first two structures are Si analogs of adamantane and diamantane, the rest are spherical nanocrystals which are build around a central atom (mostly T_d symmetry). Despite their high symmetry the spectra of the nanocrystals are predicted to become more continuous with increasing size. These predictions contrast our results of a spectrum with strong, equally spaced resonances for the highly symmetric 3D basic shape. A sudden transition from sharp, distinct features to smooth spectra, as found for diamondoids going from two to three crystal cages, is also not found for comparable silicon structures.

In conclusions, the above results show that the experimental data for the optical gaps and their interpretation on the basis of group theory have fostered the understanding of diamond nanocrystals. Further, the optical data provide a benchmark data set that have helped to improve theoretical modeling [96]. The functioning theoretical models, in turn, help to develop a deeper understanding of the experimental data. The extension of the so verified theoretical methods and their application to experimentally unaccessible systems, such as analog hydrogenated Si-nanocrystals, could provide a valid path to highly accurate theoretical predictions of the optical properties of nanoscale systems.

5.2 Photoluminescence

The discovery of intrinsic photoluminescence from Si-nanostructures in 1990 [69] caused much excitement among scientists. The concepts which are thought to underly the luminescence of nano-Si and which have been presented in section 3.1 are of general nature. Diamond in its bulk form, just like silicon, is an indirect band gap semiconductor. Diamondoids are an ideal system to investigate the luminescence of diamond nanostructures. In diamondoids, several effects which compete in typical silicon nanocrystals complicating the interpretation of photoluminescence data can be excluded *a priori*. Due to the known structure and size of diamondoids recombination at poorly defined defect sites, a broadening of the spectrum due to size distribution or Förster-type energy transfer from smaller to larger nanocrystals [100] are not present. This allows to investigate the remaining aspects of the nanocrystal luminescence, which can thus said to be *intrinsic* in nature. The term *intrinsic* is used here, despite the fact that diamondoids are *hydrogen*-passivated, to differentiate the observed luminescence from radiative transitions which occur at structural impurities giving rise to recombination sites, such as poorly defined surfaces.

The results on the photoluminescence of adamantane which are presented in this section have been published in 2009 in an article in *Physical Review B*.[9] In the following the presented data are expanded to larger diamondoids and size-dependent effects are included in the discussion of the photoluminescence of diamondoids.

The spectrally resolved photoluminescence of several diamondoids is shown in Fig. 5.16. The spectra have been scaled to similar height to facilitate the comparison. The spectra are similar in their emission energy and have in common that they are very broad in energy and without apparent sharp features. They are also relatively noisy. The smoothing fits which are shown in Fig. 5.16 together with each spectrum serve to get rid of small scale noise and to reveal larger spectral trends. They allow to distinguish spectral features within some of the spectra. However, it needs to be verified how much of the spectral structure is independent of variable external influence.

---

[9]L. Landt et al., *Intrinsic photoluminescence of adamantane in the ultraviolet spectral region*, Physical Review B **80**, 205323 (2009) - Ref. [101]

5.2. Photoluminescence

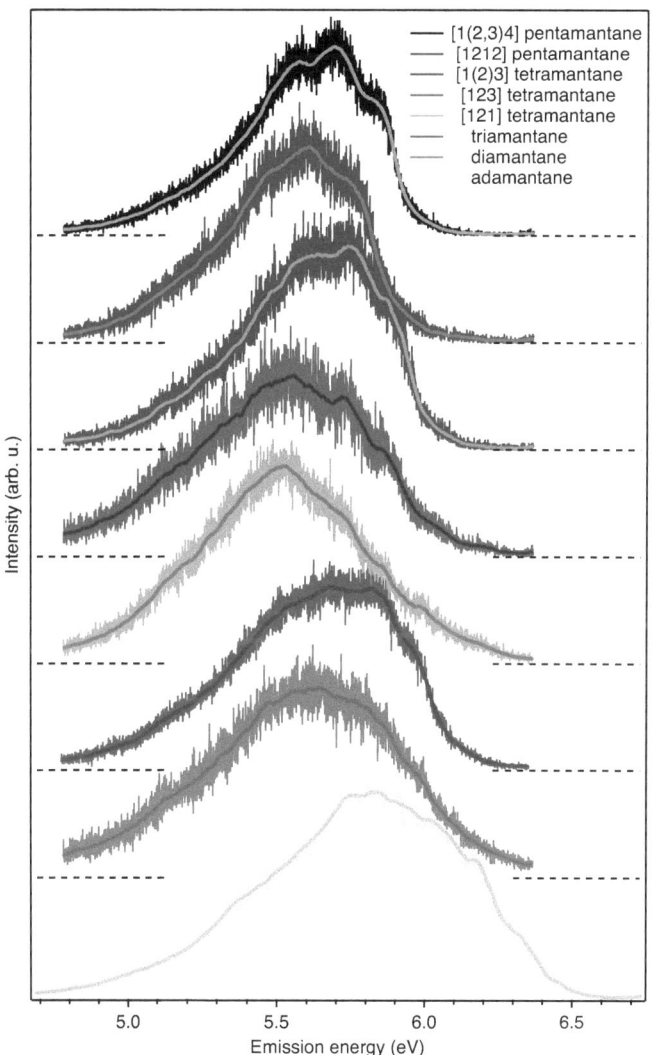

Figure 5.16: The energy resolved photoluminescence spectra of diamondoids from adamantane (bottom) to pentamantane (top) are shown. Spectra are offset and the respective baselines are indicated by dashed lines. All spectra have been scaled to similar height for sake of comparability and smoothed fits have been added to help identify spectral features.

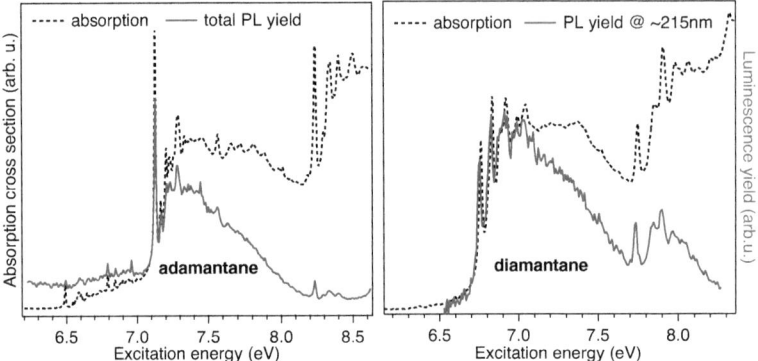

Figure 5.17: The luminescence yield of adamantane (left) and diamantane (right) compared to its optical absorption [90]. For adamantane absorption is compared to the total photoluminescence (PL) yield. For diamantane the detected luminescence light is limited to a spectral region of approx. 30 nm width centered around 215 nm. This wavelength region, corresponding energies of $\sim 5.4 - 6.2$ eV, contains the lion's share of the diamantane PL peak shown in Fig. 5.16. The differences in the absolute intensity at higher energies are due to a decreasing quantum yield of the PL.

Several checks have been carried out to make sure that the measured signal does not have its origins in contaminations, stray or scattered light, or other unforseen effects. Photoluminescence measurements are much more sensitive to contaminations than, e.g., absorption measurements because the effects in the absorption spectra are multiplied by the difference in the luminescence quantum efficiency. Differences in quantum yields can easily be on the order of 10^{-3} or 10^{-4} enhancing the effects of small contaminations by a factor of 1000-10000. It is therefore of uttermost importance to make sure that the detected photoluminescence signal originates from the investigated sample, i.e., in the present case from the diamondoids.

As a first test, the open (and thus empty) absorption cell was measured. No photoluminescence signal was obtained which means that stray light from the beamline and light that might be scattered, e.g, of the entrance slit of the absorption cell did not contribute in any significant way to the recorded spectra. Second, no significant count rate was measured below the absorption edge of the respective sample. This excludes the possibility of relevant contaminations with energy gaps smaller than those of diamondoids, such as most aromatics. The most

5.2. PHOTOLUMINESCENCE

cogent piece of evidence for the intrinsic origin of the observed photoluminescence is provided in Fig. 5.17: In the left-hand panel the integrated photoluminescence signal is plotted against the excitation energy (solid curve) and overlayed with the optical absorption of adamantane (dotted line). There is a one-to-one agreement between all characteristic features of the absorption and the luminescence signal. The growing deviation in intensity is due to a decreasing quantum yield for excitation above the band edge which will be discussed later. This investigation has been conducted in the gas phase and thus the photoluminescence can also not occur via charge transfer to impurities within the sample. The agreement of the total photoluminescence yield with the absorption spectrum therefore provides solid evidence that adamantane emits light upon photoexcitation.

Because the luminescence signal has been detected in the 0th order of the secondary monochromator it does not yield any insight as to the wavelength of the detected luminescence. Therefore one could still reasonably argue that photoluminescence from diamondoids could lie in a different, uninvestigated spectral region and be decoupled from the results presented in Fig. 5.16. To counter this objection only the photoluminescence within the spectral region of interest was measured as a function of the excitation energy. The result is shown in the right-hand panel of Fig. 5.17 using the example of diamantane. The detection area of the secondary monochromator was centered around ~5.8 eV (215 nm), such that only photons with energies between 5.4 and 6.2 eV were detected which covers the lion's share of the diamantane photoluminescence peak shown in Fig. 5.16. The luminescence yield signal in Fig. 5.17 still perfectly matches the diamondoid absorption (neglecting the deviating trend towards higher excitation energies). This provides the coherent and incontrovertible proof that the signal in Fig. 5.16 has its origins in the intrinsic photoluminescence of diamantane.

The spectra in Fig. 5.17 demonstrate the suitability of the experimental method and the correctness of the data acquisition and analysis techniques. They show, based on the data for adamantane and diamantane, that the data in Fig. 5.16 are dependable and allow for a discussion of the photoluminescence properties of diamondoids.[10]

The next endeavor, before a detailed discussion of the spectra, will be a critical assessment of the dependability of the smoothed spectra as a starting point of

---

[10]The similarity, and yet not sameness, of the data for all diamondoids further excludes the theoretical possibility of adamantane or diamantane contamination leading to the detection of photoluminescence for larger, non-luminescing diamondoid species.

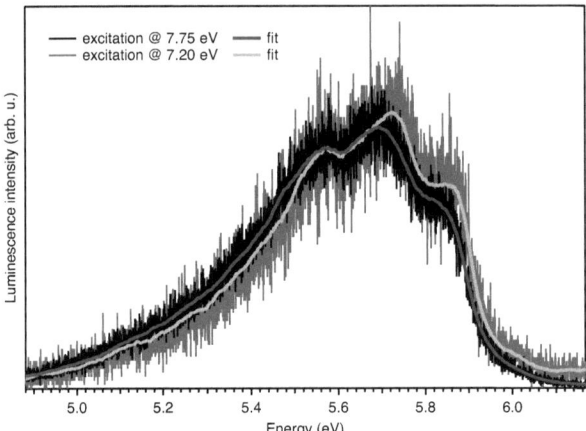

Figure 5.18: Comparison of the photoluminescence spectra of [1(2,3)4] pentamantane for excitation at 7.75 eV (160.0 nm) and at 7.20 eV (172.2 nm). The smoothed fits for both spectra reveal slight variations in the spectral shape and a shift of $\sim$ 20 meV. These differences are within the error bars of the method.

such a discussion. The smoothed photoluminescence spectra which are shown in Fig. 5.16 together with the original data vary slightly in energetic position as well as in their spectral structure. Some of them exhibit distinct spectral features. The reproducibility of the photoluminescence spectra has already been demonstrated in section 4.5. Given the noisiness of the photoluminescence signal, however, it needs to be verified that this also applies to the single spectral features which become apparent in the smoothed spectra. To put the discussion of the spectra and the conclusions drawn therefrom on solid grounds an evaluation is given in the following.

For its pronounced spectral structure [1(2,3)4] pentamantane is chosen as an example to discuss the dependence of the photoluminescence spectra on the excitation energy and the reproducibility of spectral features. In Fig. 5.18 two photoluminescence spectra of [1(2,3)4] pentamantane are shown which have been recorded at excitation energies of 7.20 eV and 7.75 eV. They have been scaled to the same height for the sake of comparability. The spectrum which was recorded at an excitation energy of 7.75 eV has the better statistics and is therefore used throughout the thesis as the standard spectrum of [1(2,3)4] pentamantane. From the comparison in Fig. 5.18 it can be seen that the spectral shape of

5.2. PHOTOLUMINESCENCE

[1(2,3)4] pentamantane is reproducible with all its features. However, the two spectra vary slightly in the relative magnitude of the features. Two aspects stand out: First, the spectrum recorded at a lower excitation energy is shifted slightly (by ~20 meV) towards higher emission energies. And second, the first two of the three main peaks in the spectrum (at approx. 5.9, 5.75, and 5.6 eV) are noticeably stronger with respect to third resonance and the low energy tail. The observed effects may be due to a dependence of the photoluminescence spectrum of [1(2,3)4] pentamantane on the excitation energy. The two spectra, however, provide insufficient data to conclude unambiguously a dependence of the spectra on the excitation energy because they differ only within the experimental error. The shift of 20 meV might simply be due to minor errors in the fairly complex data acquisition analysis process (compare section 4.5). The intensity variation of the main features within the two spectra is less than 10%. Considering the noise level of the spectra, a reproducibility within 20 meV seems to be within reasonable error limits. Thus, a dependence of the photoluminescence spectrum of diamondoids on the excitation energy can neither be confirmed nor excluded with certainty. The possible magnitude of such a dependence, however, is limited by the present data. It can be concluded from the data that

a) the photoluminescence spectra and their single spectral features are reproducible

and

b) a possible dependence of the spectra on the excitation energy is small enough to be neglected for a qualitative comparison of the presented spectra.

The above analysis shows that the smoothed fits in Fig. 5.16 provide a reliable image of the photoluminescence properties of diamondoids. A comparison of the smoothed photoluminescence spectra to the absorption sows that, in general, diamondoid photoluminescence occurs approximately from the absorption onset and downwards to lower energies. In Fig. 5.19 this is exemplified showing the comparison of the photoluminescence and the absorption spectra for adamantane (left) and [1(2,3)4] pentamantane (right). Next, the question about the physical mechanisms underlying the energetically broad luminescence arises.

Just like diamondoids many saturated hydrocarbons are known to exhibit a spectrally broad luminescence when photoexcited in the ultraviolet [102, 103]. The photoluminescence spectra of a number of different saturated hydrocarbons, taken from Ref. [102], are shown in Fig. 5.20. Comparable to diamondoids, the spectra

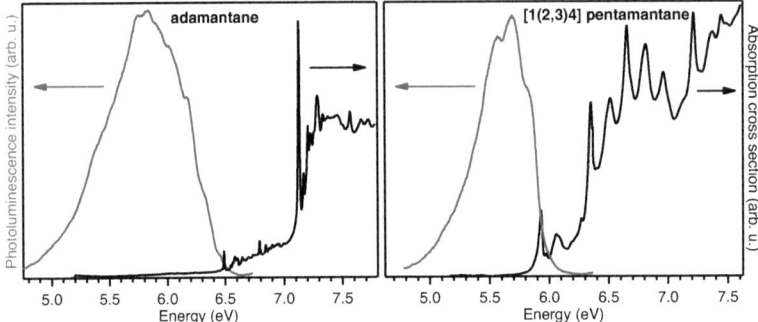

Figure 5.19: A direct comparison of the photoluminescence (red shifted, left axis) and absorption (right axis) spectra using the examples of adamantane and [1(2,3)4] pentamantane. For [1(2,3)4] pentamantane the first absorption peak coincides with a strong increase in the photoluminescence signal.

for the different structures are in general very broad in energy (FWHM ~1 eV). The most interesting of the structures in Fig. 5.20 is cyclohexane which is akin to diamondoids.[11] In Fig. 5.21 the photoluminescence spectrum of cyclohexane is directly compared to that of adamantane. The spectra have been scaled to the same height for better comparability. The cyclohexane spectrum appears shifted to higher energies by ~0.5 eV compared to the adamantane spectrum. This corresponds very well with the different absorption onsets of 6.49 eV and ~7 eV [32] for adamantane and cyclohexane, respectively. Also, both spectra are comparably broad in energy with a FWHM of 0.83 eV (adamantane) and 0.96 eV (cyclohexane). The larger width for cyclohexane, indeed, fits extremely well into the observed trend of increasing spectral width with decreasing diamondoid size (compare table 5.3 and Fig. 5.27).

The strong spectral resemblance in combination with the closely related structure implies common mechanisms behind the photoluminescence of adamantane and cyclohexane. This relation between cyclohexane and adamantane provides a starting point for the interpretation of the photoluminescence of diamondoids based on findings for saturated hydrocarbons. However, while many studies treat various aspects of the luminescence of cyclohexane [102, 103, 104, 105] none of these studies identifies the exact nature of the radiative transitions which are

---

[11] As a matter of fact, cyclohexane can justifiably be regarded as a diamondoid with polymantane order zero. Compare chapter 2.

5.2. PHOTOLUMINESCENCE

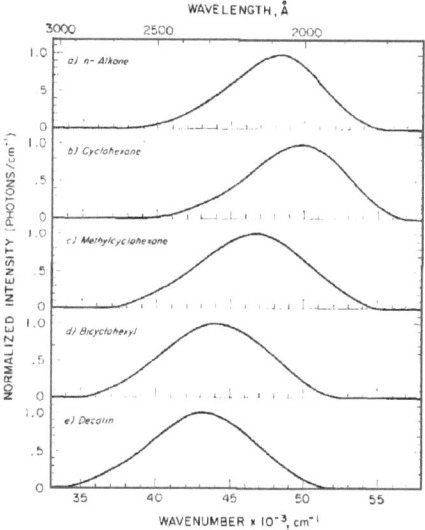

Figure 5.20: The photoluminescence spectra of some saturated hydrocarbons excited at 147 nm (8.44 eV) taken from Ref. [102]. All spectra are smooth and very broad in energy.

responsible for the luminescence.

Besides the strong similarities, however, there are also obvious differences between the photoluminescence spectra in Fig. 5.21. Most notably, the very smooth and almost gaussian shape of the cyclohexane spectrum (see inset in Fig. 5.21) is contrasted by the existence of several smaller spectral features for adamantane. The spectral structure of diamondoids clearly differentiates them from the saturated hydrocarbons shown in Fig. 5.20 and the resemblance to the smooth and very broad cyclohexane spectrum decreases as the spectra grow narrower and more feature-rich with increasing diamondoid size. The more complex, 3-dimensional structure of diamondoids appears to give rise to new mechanisms in the radiative recombination processes. A further analysis is required to identify the mechanisms underlying the photoluminescence characteristics and causing the spectral structure.

In Fig. 5.22 the smoothed photoluminescence spectra of selected diamondoids are compared. In the left panel the photoluminescence spectra of the 3D series (as introduced in section 5.1.1) are plotted next to each other. They have been scaled

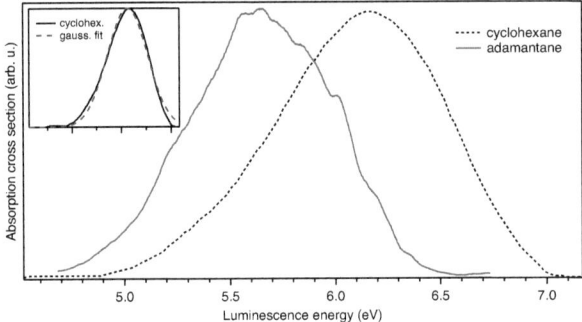

Figure 5.21: The photoluminescence spectra of adamantane compared to cyclohexane [102]. The photoluminescence of adamantane is shifted to lower energies. Both spectra are broad in energy but the adamantane spectrum exhibits slight features while the spectrum of cyclohexane is entirely smooth. The spectra have been scaled to same height for better comparability. The inset shows the spectrum of cyclohexane with a single gaussian fit.

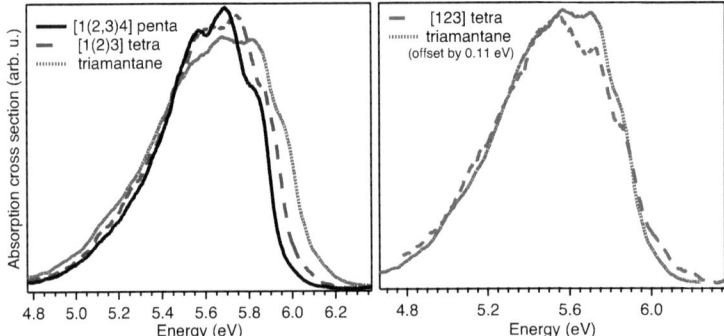

Figure 5.22: Left: Comparison of the photoluminescence spectra of the 3D family (including triamantane). The same spectral features exist for the compared diamondoids. The high energy side of the spectrum shifts to lower energies with increasing diamondoid size while the low energy tail remains fairly constant. Right: Comparison of triamantane and [123] tetramantane. The triamantane spectrum is shifted to overlay the tetramantane spectrum. Similar features can be identified yet with different relative intensities.

5.2. PHOTOLUMINESCENCE

to the same height for the sake of comparability. One can see a distinct red shift of the high energy edge with increasing diamondoid size. The low energy tail of the emission stays fairly constant in energy leading to the overall narrowing of the emission with increasing size which can also be observed in the FWHM of the photoluminescence peaks listed in Tab. 5.3. Yet, the spectral shape for all three diamondoids is very similar: The high energy flank of the emission possesses a shoulder at approximately 3/4 of the maximum intensity and the maximum exhibits a double peak. In the right panel of Fig. 5.22 the spectrum of triamantane is overlayed with the blue-shifted spectrum of [123] tetramantane. It is apparent that both spectra possess the same features. These are, however, shifted in energy and weighted differently.

Fig. 5.23 shows a fit of the smoothed photoluminescence spectrum of [1(2,3)4] pentamantane. The characteristic spectral structure with three central features can be reproduced using ten gaussian peaks for the entire spectrum. Interestingly, the best fit results in peaks with an almost equidistant energy spacing. At the high energy flank of the spectrum a very small peak at $5.982\,\text{eV}$ is followed by the three main peaks at 5.840, 5.700, and $5.566\,\text{eV}$. The energy spacing between these first four peaks is fairly constant at $\sim 140\,\text{meV}$. The subsequent six peaks follow with a constant spacing of $\sim 120\,\text{meV}$ each. A constant energy spacing is typical for vibronic energy levels. In fact, the constant spacing is valid for a harmonic potential which is only a good approximation for the low-lying states. Qualitatively, the spacing decreases for higher lying vibrational states, which is what is observed for the single peaks of the fit. This decomposition of the spectrum thus hints towards phonon-assisted transitions being responsible for the shape of the photoluminescence spectrum. Similar gaussian fits have been used to identify phonon-assisted transitions in Si nanocrystals [106]. Prominent Raman modes for [1(2,3)4] pentamantane exist at 147 and $133\,\text{meV}$ [27].

Radiative transitions between two electronic levels occur according to the Franck-Condon principle involving states which have different degrees of vibrational excitation. The transition probabilities between the different vibronic levels depend on the Franck-Condon factors, i.e., the overlap of the wave functions of two states in a given geometry. In Fig. 5.24 the potentials of a system in the ground state and in the excited state are shown schematically over the configuration variable R, i.e., basically over a parameter set for the internuclear distances. According to the Franck-Condon principle transitions between the two states occur vertically, i.e., at constant atomic coordinates. A tentative assignment of the transitions in

CHAPTER 5. RESULTS & DISCUSSION - PART I: PRISTINE DIAMONDOIDS

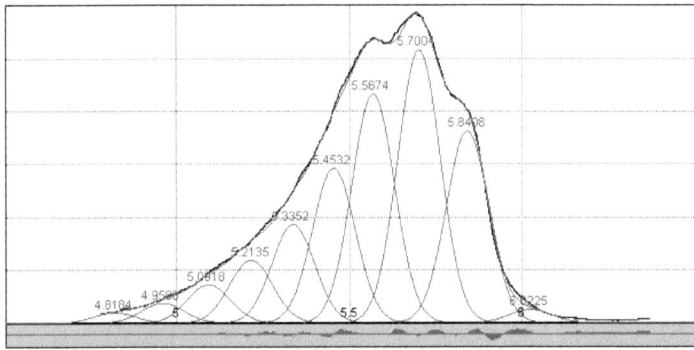

Figure 5.23: A fit of the photoluminescence spectrum of [1(2,3)4] pentamantane. Ten gaussians were necessary to fit the spectrum; the residual is shown below the fit. The three most prominent peaks which lend the spectrum its characteristic shape possess a spacing of approx. 140 meV, the peaks forming the tail exhibit a spacing of ~120 meV.

[1(2,3)4] pentamantane which are identified in the fit in Fig. 5.23 is attempted in Fig. 5.24. The transitions between different vibrational states of the excited state and the ground state potentials are indicated by vertical lines. The length of the line signifies the energy of the emitted photon. The Franck-Condon factors, i.e., the intensity of the corresponding transitions, are indicated by a color-scale: Franck-Condon factors range from large (black, strong transition) to small (light-grey, weak transition).

The fact that the Franck-Condon factors are non-zero for nine different transitions indicates a large distortion of the diamondoid in the excited state. This distortion is reflected in a large shift of the excited state potential in Fig. 5.24 which is required to obtain nine different vertical transitions. Further, large Franck-Condon factors imply a strong electron-phonon coupling [107] which typically gives rise to a high rate of non-radiative decay due to internal conversion. The radiative and non-radiative rates will be discussed later. The first weak transition which is the highest in energy is not likely to be due to the $0 \leftarrow 0$ transition as it appears at 5.98 eV and thus shifted towards higher energies with respect to the optical gap of [1(2,3)4] pentamantane. Thus, it seems reasonable to assume that the first of the three main resonances at ~5.9 eV represents the $0 \leftarrow 0$ transition and the following peaks the $0 \leftarrow 1, 2, 3, \ldots$ transitions. This assignment, however, is not the only viable one. It means that the $0 \leftarrow 0$ occurs at a central energy of 5.84 eV. Compared to the center of the first absorption peak at 5.93 eV this

5.2. Photoluminescence

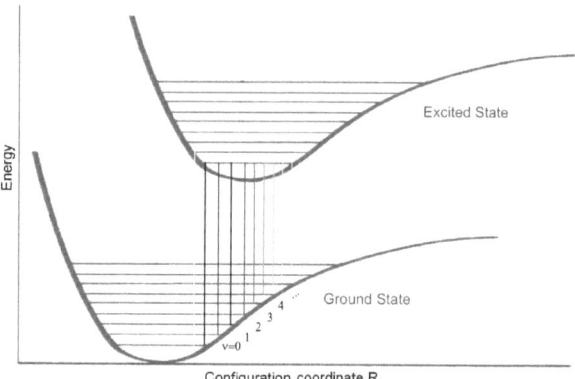

Figure 5.24: According to the Franck-Condon principle transitions between two different electronic states occur under fixed nuclear coordinates. i.e., vertically in this scheme. Due to a distortion of the system in the excited state the potential surface shifts. Here, the components of the spectrum [1(2,3)4] pentamantane as identified in Fig. 5.23 are tentatively assigned to vibronic transitions between the excited and the ground state. The Franck-Condon factors are indicated by a color-scale: black transitions have large and light-grey transitions have small Franck-Condon-Factors. Note that in this assignment the $0 \leftarrow 1$ transition is the highest in energy.

results in a Stokes shift of approximately 0.1 eV. Note that in the assignment of the transitions in Fig. 5.24 the weak transition at the high energy edge of the photoluminescence is due to the $0 \leftarrow 1$ transition. Such a peak occurs to the right of the $0 \leftarrow 0$ transition in Fig. 5.23 and is due to a relaxation from a vibronically excited state. This means that a phonon is annihilated together with the recombination of the electronically excited state to contribute to the energy of the emitted photon. At moderate temperatures this process is unlikely compared to the inverse process where with the recombination process a phonon is created. This assignment is in agreement with the large difference in the intensities of the neighboring peaks.

In Fig. 5.25 fits for some other diamondoid structures are shown. While it is always possible to approximately fit the spectrum using 10 − 12 gaussians, depending on the spectral width of the emission curve, the position of the single

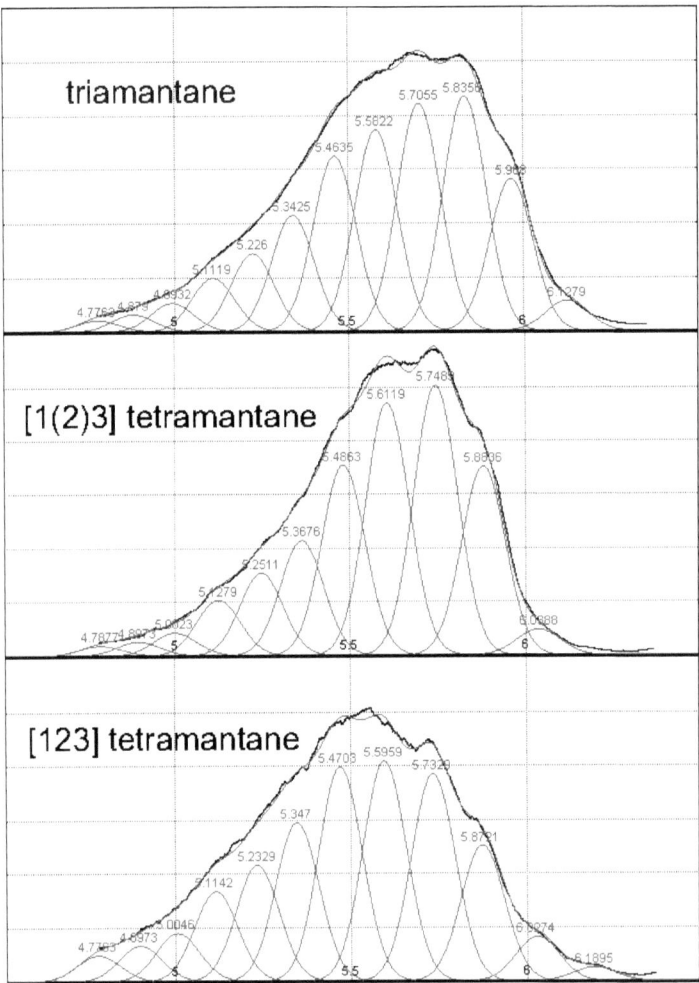

Figure 5.25: Fits of the photoluminescence spectra of triamantane, [1(2)3] tetramantane, and [123] tetramantane. The spacing between the single components is $\sim 120-140\,\text{meV}$, similar to [1(2,3)4] pentamantane.

5.2. PHOTOLUMINESCENCE

peaks becomes less obvious with vanishing spectral structure. For triamantane, [1(2)3] tetramantane, and [123] tetramantane shown in Fig. 5.25 an unequivocal assignment of the first, high-energy peaks is still possible. But already for the third of the larger peaks the exact energetic position is contentious as can be seen from the deviation between fit (black curve) and spectrum. For smaller diamondoids with even fewer spectral structure a fit of the spectra is still feasible. An unambiguous assignment of single transitions, however, is no longer possible.

The tentative assignment for the radiative transitions of [1(2, 3)4] pentamantane allows to identify single transitions and to estimate the relative magnitudes of their Franck-Condon factors. However, for most diamondoids the model is not able to unambiguously determine the energetic positions of the involved transitions. But a generalization of the underlying concepts provides a viable explanation for some of the observed features of the diamondoid photoluminescence, such as the unusual spectral width of the emission and the asymmetric peak shape. Since the contribution of phonon-assisted transitions changes with temperature, temperature-dependent measurements of the photoluminescence spectrum could provide the means to experimentally verify the validity of the model.

The above analysis of the energetically broad emission of diamondoids indicates a severe distortion of the diamondoid framework in the excited state. This strong distortion occurs despite a large structural rigidity of the diamondoid cage structure. A phenomenon which could be responsible for the observed lattice distortions and the resulting luminescence in diamondoids is the formation of so-called self-trapped excitons. The self-trapping of an exciton can occur when a strong distortion of the nuclear framework leads to a local minimum in the potential curve that is offset with respect to the global minimum, i.e., the fully relaxed excited-state geometry. Such self-trapped excitons have been observed in several nanoscale systems and are expected to occur in all kinds of semiconductor nanocrystals [108]. The principle of self-trapped excitons is sketched in Fig. 5.26: The recombination of a self-trapped exciton (STE) occurs at nuclear coordinates which are far from the equilibrium geometry of the ground state. The trapping potential that is offset in the atomic configuration (x-axis) with respect to the excited state is typically due to a local distortion, e.g., due to an excitation which is localized on a single bonds of the particle. The local minimum in the potential surface leads to the trapping of the exciton and prevents a relaxation to the lowest energy geometry of the excited state. As a consequence, two things may occur:

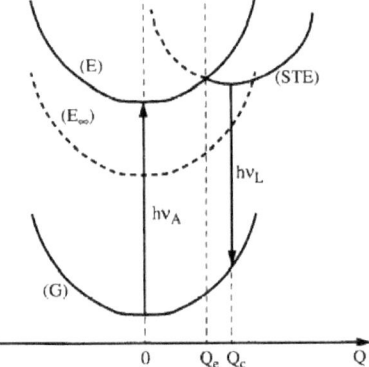

Figure 5.26: Diagram of the self-trapped exciton model [108]. An electron is promoted from the ground state (G) into the excited state (E). The electron can also get trapped in a potential that possesses a local minimum. Such potentials which can arise due to local distortions lead to self-trapped excitons (STE) and can explain energetically broad and strongly red-shifted luminescence properties.

i) the electron eventually tunnels through or hops over the potential barrier and the system relaxes into the regular excited state geometry from where it relaxes into the ground state, or ii) the system relaxes into its electronic ground state directly from the self-trapped exciton geometry. Obviously, case i) is not relevant because it cannot be told apart from regular radiative decay. Case ii) on the other hand gives rise to a significant red shift and a broadening of the luminescence peak. This case can be expected to dominate if recombination of the self-trapped exciton is fast compared to the relaxation to the excited state geometry.

The fact that an assignment of single transitions in the photoluminescence peaks is not possible for all diamondoids, makes an unambiguous determination of characteristic numbers for the emission energy or the Stokes shift difficult. Thus, in the following several spectral parameters are defined which are thought to best characterize the properties of diamondoid luminescence and which shall enable a quantitative discussion.

The spectral width of the emission is quantified by its full-width half-maximum at the half of its maximum intensity (FWHM). Besides serving as a measure for the spectral width of the emission it is further used to determine the central

5.2. PHOTOLUMINESCENCE

diamondoid	PL_{min}	PL_{max}	FWHM	E_{exc}	PL center	opt. gap	Stokes
adamantane	4.96	6.43	0.83	7.12	5.82	6.49	0.67
diamantane	4.77	6.29	0.78	7.90	5.61	6.40	0.79
triamantane	4.91	6.15	0.69	7.43	5.66	6.06	0.40
[121]-tetram.	4.84	6.19	0.66	7.75	5.51	6.10	0.59
[123]-tetram.	4.75	6.10	0.73	7.90	5.52	5.95	0.42
[1(2)3]-tetram.	4.98	6.04	0.58	7.75	5.64	5.94	0.30
[1212]-pentam.	4.94	5.97	0.53	7.87	5.56	5.85	0.28
[1(2,3)4]-pent.	5.01	5.88	0.51	7.75	5.62	5.81	0.19

Table 5.3: Characteristic values for the photoluminescence of diamondoids. All values are in eV. PL_{max}/PL_{min} describe the upper/lower bound where the photoluminescence signal reaches 10% of its maximum value. The 'central PL' gives the center of the FWHM-region and is used as a measure of the spectral shift with diamondoid size. The 'Stokes' shift reflects the difference between the optical gap and the central PL energy for each diamondoid. The excitation energies are listed under E_{exc}.

emission energy (or central photoluminescence energy) which is simply defined as the arithmetic mean of the two energies where the emission intensity reaches half its maximum value. Further, a *region of photoluminescence* has been defined as the spectral region where the luminescence intensity exceeds 10% of its peak value. The upper and lower bounds of this region are labeled PL_{max} and PL_{min}, respectively.

In Fig. 5.27 these characteristic numbers are used to visualize the photoluminescence properties as a function of diamondoid size. The region of photoluminescence (PL) is shaded in light grey. The energy where the photoluminescence intensity reaches 50% of the peak value is marked by two dashed lines and the region inside these boundaries is shaded in darker grey. The arithmetic center of the FWHM region is marked by solid dots for each diamondoid that are connected by a dashed line to visualize the trend. Additionally, the optical gaps are shown in the form of black solid dots also connected by a dashed line to guide the eye. Note that the three tetramantanes and the two pentamantanes have been slightly offset with respect to each other to enable graphical evaluation. The sequence of diamondoids in the graph is the same as the order of diamondoid names listed in Tab. 5.3. Changing this sequence does not, however, in any way affect the interpretation of the graph nor the conclusions drawn from it. The characteristic numbers for the photoluminescence of each diamondoid are listed together with

CHAPTER 5. RESULTS & DISCUSSION - PART I: PRISTINE DIAMONDOIDS

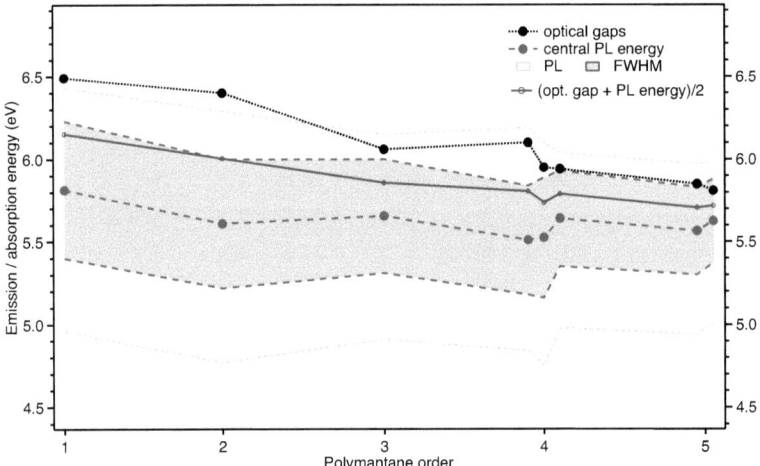

Figure 5.27: The region of photoluminescence (PL) is shown including full-width half-maximum (FWHM) and the center of the luminescence peak. For comparison the optical gaps determined in section 5.1.3 are plotted as well.

the optical gaps from section 5.1.3 in Tab. 5.3.

Fig. 5.27 illustrates that emission occurs over a wide spectral range for all diamondoids and that emission energies, and especially the low energy limit of the photoluminescence, are fairly constant. A discernible trend is the narrowing of photoluminescence peaks with growing diamondoid size. The FWHM starts out with 0.83 eV for adamantane and reaches its minimum for [1212] pentamantane with 0.53 eV, i.e., 0.3 eV less. Also a very subtle trend of decreasing emission energies with increasing diamondoid size is observed. The overall shift of the central photoluminescence energy just barely reaches 0.2 eV from adamantane to the pentamantanes. This compares to a shift in the optical gaps of almost 0.7 eV over the same size range. Comparing luminescence energies to the optical gaps and the shift between them is somewhat bothersome because different key figures for the photoluminescence from diamondoids exist. Further specification and a definition of representative values comparable to a Stokes shift are required.

The Stokes shift is usually defined as the difference in the energy of the absorption edge and the emission energy. In the case of diamondoids it is impossible to reasonably define one emission wavelength and thus to derive a Stokes shift in the classical sense. Here, alternatively, the Stokes shift, which is also listed

5.2. PHOTOLUMINESCENCE

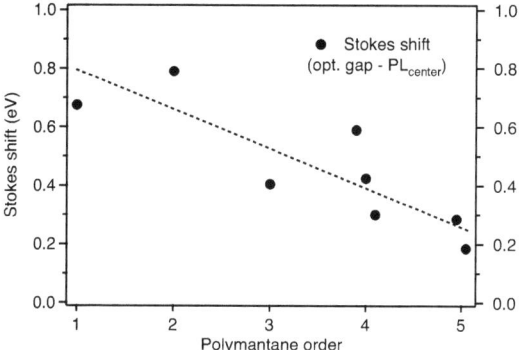

Figure 5.28: The Stokes shift as a function of diamondoid size. The Stokes shift is here defined as the difference between optical gap and the center of the photoluminescence.

in Tab. 5.3, is defined as the difference between the optical gap and the central photoluminescence energy. The so-defined Stokes shift is plotted over the diamondoids' polymantane order in Fig. 5.28.

The Stokes shift exhibits a clear size-dependence which means that for increasing diamondoid size the energy difference between absorption and emission center diminishes. This is visualized by the addition of a finely dashed trend line for the black dots. Values range from 0.8 eV for diamantane to 0.2 eV for [1(2,3)4] pentamantane. Comparison with Fig. 5.27 shows that the trend is in large parts due to the strongly size-dependent optical gap. An interesting observation for the absorption and emission properties of diamondoids in Fig. 5.27 is the fact that, when deviations from the respective trends for absorption and luminescence occur, they seem to happen in opposite directions. This means that when the optical gap of a structure is comparably high in energy with respect to the overall trend, its photoluminescence peak is comparably low in energy. This curious observation is best visualized by the gray solid line in Fig. 5.27 which marks the arithmetic mean of optical gap and the central photoluminescence energy. This line is nearly flat indicating that deviations from a size-dependent trend occur not only in opposite directions for absorption and emission but also at comparable magnitudes.

In the following it is attempted to derive an estimate of the photoluminescence quantum yield of adamantane. A strong wavelength dependence of the quantum yield is apparent from Fig. 5.17. Similar to the case of cyclic alkanes the quantum yield drops significantly for increasing excitation energies [105]. Thus, the quantum efficiency of adamantane should have its maximum directly at the absorption edge of 6.5 eV. For reasons of signal intensity, however, an excitation energy of 7.12 eV (174.2 nm) is chosen.

The photoluminescence quantum yield (ϕ) of a system is given as the ratio of the number of emitted photons to that of absorbed photons:

$$\phi = \frac{N_{emis}}{N_{abs}}$$

As a first step the number of absorbed photons is estimated: The number of absorbed photons N_{abs} is the difference between the number of transmitted photons in the empty and in the filled absorption cell. These numbers (or better the rates, i.e., the photons/s) can be estimated from the measured current of the photodiode:

$$N_{abs}/s^{-1} = \frac{I_{full} - I_{empty}}{e} \cdot QE_{diode}$$

Here, $I_{full/empty}$ is the diode current measured for the full/empty absorption cell, e is the charge of the electron, and QE_{diode} is the quantum efficiency of the diode. At the excitation wavelength of 174 nm the data sheet of the used GaP photodiode shows a quantum efficiency of 1%.

As indicated in the schematic drawing in Fig. 5.29, not the absorption over the entire length of the absorption cell is relevant for the calculation. Only light emitted from approximately the central third of the absorption length can escape through the slit onto the detector and thus contribute to the signal. Therefore, N_{abs} has to be limited to the number of photons which are absorbed on the second third l_2 of the entire absorption path. In the case of adamantane, the absorption over the entire length of $L = 48$ mm is measured to be 60 % at 174.2 nm excitation. This allows to determine the absorption coefficient α according to the Lambert-Beer law

$$N_{abs} = N_0(1 - e^{-\alpha L}) = 0.6 \cdot N_0$$

Solving this equation for α gives an absorption coefficient of $\alpha = 0.019$ mm^{-1} for the present experimental conditions. The absorption in the relevant area of the absorption cell is the difference of the intensity at the beginning and at the end of this segment

$$N_{abs}(l_2) = N_1 - N_2 = N_0 \cdot (e^{-\alpha l} - e^{-\alpha 2l})$$

and using the value for α determined from the total absorption the absorption within the segment l_2 is 19.2%. The light transmitted through the empty cell

5.2. PHOTOLUMINESCENCE

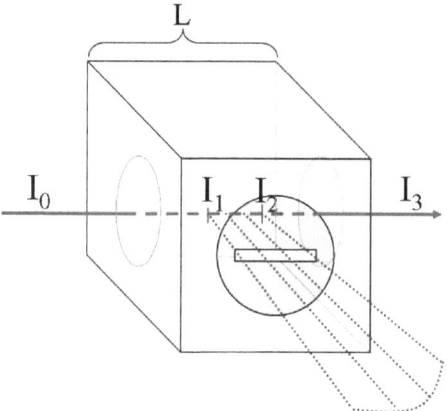

Figure 5.29: Schematic drawing of the absorption (dashed line) and the relevant photoluminescence (dotted lines). The relevant photoluminescence is the light that is emitted such that it escapes through the exit slit of the cell window in such an angle that it hits the detector. This segment is confined to approximately the second third of the absorption path. The intensities I_0, and $I_{1/2/3}$ are the initial intensity of the monochromatized synchrotron beam and the Intensities after 1, 2, and 3 thirds of the absorption path through the cell, respectively.

caused under the prevailing conditions a diode current of $I_{empty} = 2.82\,\text{nA}$. Considering the quantum efficiency of the diode this turns out $1.8 \cdot 10^{12}$ photons / s and $N_{abs}(l_2) = 3.4 \cdot 10^{11}$ photons / s.

Next, one needs to calculate the numbers of photons emitted per second from the diamondoid sample. The emission characteristics of the diamondoids in the gas phase is uniform, i.e., there is no preferred direction. This means that of all emitted photons only a fraction corresponding to the solid angle of the detector is headed the right way. Further efficiencies of the optics and the detector have to be accounted for. The calculation of the emitted photons is then straight forward:

$$N_{emis} = N_{det} \cdot \frac{1}{\Omega_{det}} \frac{1}{QE_{det}} \frac{1}{R_{mirror}} \qquad (5.1)$$

N_{det} is the number of detected photons, Ω_{det} the solid angle of the detection area, QE_{det} the quantum efficiency at the detected wavelength and R_{mirror} is the reflectivity of the mirror in zero-order. The quantum efficiency of the CsTe-detector over the wavelength region of 200-240 nm is estimated to be 30% and the reflectivity of the mirror at those wavelength is also approximately 30% [86].

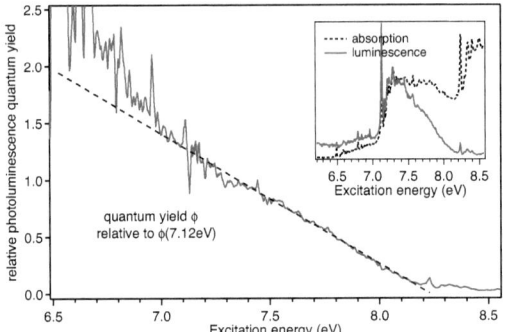

Figure 5.30: Dependence of the quantum yield of adamantane on the excitation energy. The curve is the photoluminescence signal divided by the absorption signal. It is normalized to the quantum yield at 7.12 eV (174.2 nm). It shows an almost linear dependence of the quantum yield with the excitation energy. The deviation at lower energies is due to background in the luminescence signal (compare inset).

The focussing grating at 1 m distance has an edge length of 10 cm. The slit which had to be used to suppress reflections measured $1 \times 14\,\text{mm}^2$ and limited the transmitted light only in the vertical plane. The setup detected at a solid angle of $2.8 \cdot 10^{-4}$. The number of detected photons can simply be read out from the total luminescence yield: $N_{det} = 1.45 \cdot 10^4/5\,s \simeq 3 \cdot 10^3/s$. Taken all together this allows to calculate the total number of emitted photons:

$$N_{emis} = 3 \cdot 10^3 \cdot \frac{1}{2.8 \cdot 10^{-4}} \frac{1}{0.3} \frac{1}{0.3} = 1.2 \cdot 10^8\, photons/s$$

Divided by the number of absorbed photons this provides a rough estimate for the lower limit of the quantum yield of adamantane at an excitation wavelength of 174.2 nm:

$$\phi_{ada}(\lambda = 174.2\,nm) \approx 0.04\%$$

In Fig. 5.30 the dependence of the quantum yield on the excitation energy is shown. The ratio of the photoluminescence to the absorption signal is normalized to the value at 7.12 eV (174.2 nm). A linear dependence is found for the quantum yield. The deviation at lower energies is due to background in the luminescence signal (compare inset) and varies strongly with different background subtractions. The linear trend for excitation energies from 7 to 8 eV, however, is stable and not greatly affected by different background subtractions. The dashed

5.2. PHOTOLUMINESCENCE

trend line allows to extrapolate the photoluminescence quantum yield for excitation at the absorption edge of adamantane (6.49 eV). It can be assumed to be roughly twice the value found for excitation at 7.12 eV: $\phi_{ada}(6.49\,eV) \approx 0.1\%$.
The experimental setup was not optimized to determine the quantum yield of diamondoid photoluminescence. The above numbers are therefore to be understood as a rough estimate. A dependence of the quantum yield on the size and structure has not been investigated. The photoluminescence quantum yield of diamondoids is thus, very roughly, on the order of 10^{-3} but may vary among different diamondoid structures.

The approximate number for the quantum yield allows to give an estimate for the radiative and non-radiative rates [109]. The rate of photoluminescence decay is given by the sum of the radiative and non-radiative recombination rates: $1/\tau_{PL} = 1/\tau_R + 1/\tau_{NR}$. The quantum yield ϕ, which is proportional to the photoluminescence intensity, can be estimated as the weight of the radiative channel in the recombination process: $\phi = \tau_R^{-1}/\tau_{PL}^{-1}$. The low luminescence quantum efficiency observed for diamondoids means that non-radiative decay channels are fast compared to radiative decay, or $\tau_R \ll \tau_{NR}$. Therefore, $\tau_{PL} \simeq \tau_{NR}$ and thus $\phi \simeq \tau_{NR}/\tau_R$. For adamantane, the non-radiative decay processes are thus approximately 1000 times faster than the radiative processes.

Because no structural deficiencies exist, non-radiative decay channels in diamondoids are limited to internal conversion and intersystem crossing processes. The fits which have been conducted above for several diamondoid spectra indicate a strong electron-phonon coupling in diamondoids. This observation fits well to the fairly low photoluminescence quantum yield. For increasing diamondoid size the distortion of the excited state geometry is likely to diminish due to increasing mechanical resistance of the surrounding diamondoid framework. This will lead to smaller Franck-Condon factors and in consequence to a lower non-radiative rate. It should therefore entail a trend towards increasing quantum efficiencies with larger diamondoid size. The shape of the diamondoid is also likely to play a role. Following the argumentation that higher rigidity will lead to smaller lattice distortion and, in consequence, to smaller non-radiative rates, a compact structure bears advantages over a frail structure. This interpretation is backed by the listed values for FWHM of the three tetramantane structures in table 5.3. [1(2)3] tetramantane has clearly the narrowest emission which indicates smaller Franck-Condon factors. It can therefore be expected to have the highest quantum yield of the tetramantanes. Again, as for the absorption properties, shape may play a non-negligible role for the emission behavior of diamondoids.

Chapter 6

Results & Discussion - Part II: Modified Diamondoids

One central goal of this work is to understand the optical and electronic properties of diamondoids which have been treated in the previous chapter. A second central aim of this thesis is to reveal the changes which are induced in the electronic structure and the optical properties of a diamondoid by structural modification. The influence of such modification is the topic of this chapter.

Two fundamentally different ways to chemically alter a diamondoid exist: Either atoms can be attached to the surface in form of a functional group or carbon atoms from the diamond framework can be replaced by different atoms.

First, the influence of functionalization is investigated using the example of diamondoid thiols [58]. From a fundamental perspective, the investigation of diamondoid thiols allows to determine the precise influence that a single functional group (here: the thiol group) has on the electronic structure of a small nanocrystal (here: a diamondoid). This, however, could in principle be achieved by using any of the many functionalizations that are available for diamondoids. Thiols have been chosen in this work because they are among the most common functional groups and offer various possibilities for the preparation of new hybrid materials. Thiol groups link, e.g., to noble metals and in solution they readily self-assemble on Au-, Ag-, and Pt-surfaces [49]. In the case of diamondoids the formation of self-assembled monolayers (SAMs) of diamondoid thiols on noble metal surfaces bears great prospects for nanotechnology. In particular, their use as low field electron emitters, which has been demonstrated recently [2], seems promising. Thus, for fundamental as well as for applied interest diamondoid thiols are object of investigation in this chapter. In section 6.1 a full electronic structure investigation of adamantane-1-thiol is undertaken to give a complete picture of the changes induced in the electronic structure of adamantane by the attachment of a thiol group. The investigation of the optical properties is then, in section 6.2, expanded to larger thiols.

The influence of the second approach, the bulk-substitution of single atoms, is investigated in section 6.3. Understanding the effects of single impurity atoms incorporated into nanocrystal structures is of enormous scientific and technological relevance. The species investigated in section 6.3, oxadiamondoids and urotropine, are model systems of minuscule, well-defined nanocrystals in which a single or multiple atoms are substituted by impurity atoms in a controlled manner.

6.1 Adamantane-1-Thiol - The Whole Picture

In this first section of the chapter on modified diamondoids, a complete picture of the influence of a single thiol group on the electronic and optical properties of adamantane, the simplest diamondoid, is given. For this purpose, several spectroscopic techniques have been applied to adamantane-1-thiol. The investigation features gas phase measurements of the core levels, the valence and the conduction states, a Raman analysis of the vibrational modes, and optical absorption and photoluminescence spectroscopy measurements. A comparison of the data to previous investigations on adamantane allows to isolate the influence of a single thiol functional group on each of the individual properties of the diamondoid.

Besides the fundamental interest, adamantane-1-thiol is also of technological relevance. SAMs of adamantane-1-thiol on Au(111) surfaces have been studied extensively [110, 111, 112]. They are less faulty than alkane SAMs and have, among others, been proposed as placeholder in self-assembly processes [111]. Similar structures involving an adamantane cage unit and a thiol link have also been investigated [113, 114, 115].

The findings of this section have been summarized in early 2010 in an article in *The Journal of Chemical Physics*.[1]

---

[1]L. Landt et al., *The influence of a single thiol group on the electronic and optical properties of the smallest diamondoid adamantane*, The Journal of Chemical Physics **132**, 024710 (2010) - Ref. [116]

6.1. ADAMANTANE-1-THIOL - THE WHOLE PICTURE

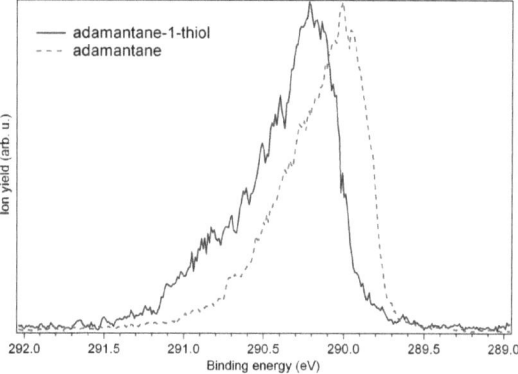

Figure 6.1: C1s core level spectra for adamantane-1-thiol and adamantane. Electrons from the C1s core levels are more strongly bound in the thiol by ∼0.2 eV. Spectra have been scaled to same height to simplify comparison.

In Fig. 6.1 the spectra of the carbon 1s core level for adamantane-1-thiol and adamantane are shown. Both spectra exhibit a steep incline on the lower binding energy side and a tail towards higher energies. While the appearance of the spectra is the same, the C1s levels of the thiol are shifted by 0.2 eV to higher binding energies. This shift is due to the transfer of electron density from the parent molecule to the highly electronegative thiol group. The electron transfer results in lower electron density in the parent cluster and thus in more strongly bound electrons. Both spectra have been scaled to the same height for the sake of comparability and the structure in the high energy tail of the adamantane-1-thiol is due to lower count rates, i.e., increased signal-to-noise ratio.

In Fig. 6.2 the photoelectron spectra of adamantane (bottom) and adamantane-1-thiol (top) are displayed. For the discussion of the data it proves useful to split the thiol spectrum in two parts, a band structure of electronic states starting at binding energies of 9.5 eV and a broad peak centered around 8.8 eV. These two parts are separated in the spectrum by a gap of a few hundred meV (located around 9.2 eV binding energy) where no electronic levels are present.
The higher energy part of the thiol spectrum beyond binding energies of 9.3 eV can be superimposed on the adamantane spectrum. These electronic states which be-

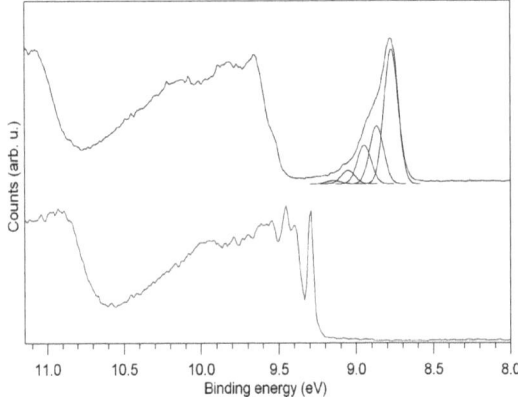

Figure 6.2: Photoelectron spectra for adamantane-1-thiol (top) and adamantane (bottom). The spectra have been scaled to same height to simplify comparison. The thiol spectrum contains the adamantane spectrum, even though it appears shifted to higher binding energies by about 0.2 eV and the vibrational structure is attenuated. An additional density-of-states arises for the thiol around 8.8 eV. It lowers the ionization potential by 0.58 eV compared to the unmodified cluster [25]. The peak has been fitted with five gaussian peaks to resolve vibronic structure.

long to the diamondoid parent molecule are not greatly affected by the thiolation. Only the resonances at the absorption edge of the adamantane spectrum which have been identified as vibronic in nature [33] are attenuated in the adamantane thiol. This part of the thiol spectrum is shifted to higher binding energy by approximately 0.2 eV. This is the same effect that is observed in the core level spectra in Fig. 6.1 which is due to redistribution of electron density upon thiol attachment and which, apparently, is similar in magnitude. The electron depletion in the diamondoid part leads to the formation of a local positive charge (polarization) thus increasing the binding energy. The good agreement with the adamantane spectrum is not self-evident because the thiolation breaks the symmetry of the cluster from T_d for adamantane to C_s for adamantane-1-thiol. This revokes the symmetry-induced degeneracy of the electronic levels found in adamantane as is obvious for the HOMO in Fig. 6.3. In fact, the HOMO in adamantane is triply degenerate with the HOMO-1 and HOMO-2 which have the same energy and all transform as T_2. The resemblance of the spectral features and of the orbitals in adamantane-1-thiol implies that the orbitals remain quasi-degenerate upon thio-

6.1. ADAMANTANE-1-THIOL - THE WHOLE PICTURE

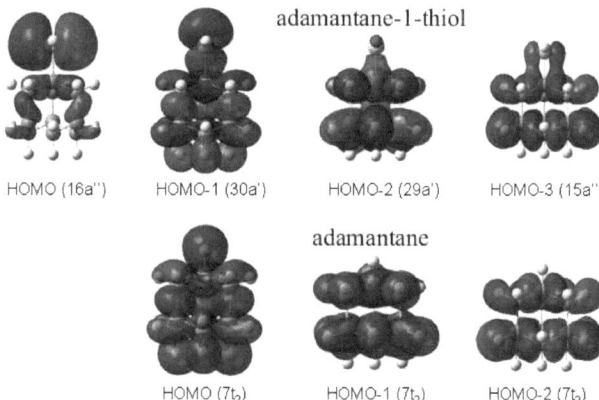

Figure 6.3: Molecular orbitals of adamantane and adamantane-1-thiol calculated at B3LYP/6-31G* level of theory as implemented in GAUSSIAN03. The HOMO of adamantane-1-thiol is given by the electron lone-pair of the thiol group, the following orbitals are fundamentally those of adamantane.

lation despite a lower symmetry.

By contrast, the lower energy part of the thiol spectrum, the broad peak centered around 8.8 eV, possesses no analog in the adamantane spectrum. By analogy with closely related molecules, such as alkane thiols [117, 118], this additional density of states can be ascribed to the $n_S^\perp$ nonbonding out-of-plane electron lone-pair from the thiol group. This is also in good agreement with the calculated orbital shape of the HOMO in Fig. 6.3. Due to this *"thiol state"* the ionization potential is lowered by 0.58 eV compared to adamantane [25] to 8.65 eV. Listed in Tab. 6.1 are the adiabatic ionization potentials which are determined as intersection of a tangent to the graph and the baseline [25]. The quantum chemical calculations of the molecular orbitals shown in Fig. 6.3 confirm that the thiol state, despite its broadness, is indeed due to only a single electronic level. The thiol state is broadened by vibronic transitions which are resolved in the fit in Fig. 6.2 using five gaussian peaks. These peaks have an energy spacing of approximately 90 meV. The predominance of the first gaussian peak at the low energy side of the thiol state suggests that the adiabatic transition is by far the strongest. This is in accordance with the non-bonding character of the orbital because electronic excitation from a bonding orbital typically leads to a strong coupling to vibronic modes which then dominate the spectrum. To identify the excited vibrational modes Raman spectroscopy was employed. The results are presented in the following section.

Fig. 6.4 compares the measured Raman spectrum of adamantane-1-thiol (a) to the spectra computed for the neutral cluster (b) and the cation (c). The Raman spectra of adamantane and adamantane-1-thiol were taken using a Dilor LABRAM spectrometer (1800 l/mm grating). The ready-to-use setup reproduced the known spectrum of adamantane [27] which also served for calibration purposes. The 488 nm line of an Argon$^+$ laser was used as exciting radiation and the spectral resolution was $2\,\mathrm{cm}^{-1}$. As the only of the techniques used in this work, Raman spectroscopy was performed on condensed samples under atmosphere.[2].

The Raman spectrum of the cation has been computed because the PES data that we strive to explain reflect the ionic final state of the cluster. This means that the data in Fig. 6.2, which will be discussed at the end of this section, contain the vibrational modes of the cluster ion. An investigation of methane thiol [117] showed that the Raman modes in the ionic geometry yields far better agreement with experimental data than computing the ion Raman modes in the ground state geometry. Therefore the optimized geometry of the adamantane-1-thiol cation has been computed and its Raman spectrum has been determined. The Raman spectrum of adamantane has been added as dotted line to Fig. 6.4 (a) for comparison.

There is good agreement between the measured (a) and the computed Raman spectrum (b) for adamantane-1-thiol. In the wavenumber region below $1000\,\mathrm{cm}^{-1}$ differences in frequencies generally remain within a few cm^{-1}. Based on the agreement of the measured and the computed spectrum of the neutral thiol cluster the computational method is expected to yield similarly reliable results for the Raman spectrum of the adamantane-1-thiol cation. The vibrational modes that have been assigned in adamantane [27] can be identified with one of the calculated Raman modes for adamantane-1-thiol. This is achieved by comparing the vibrational motion of the computed modes which can be displayed using GaussView. Most of the vibrational modes in adamantane are not severely affected by thiolation. The symmetric CC stretch (breathing) mode of adamantane which is found at $757\,\mathrm{cm}^{-1}$, e.g., is found for adamantane-1-thiol at $776\,\mathrm{cm}^{-1}$ in the measured and $770\,\mathrm{cm}^{-1}$ in the computed Raman spectrum. The same holds true for most

---

[2]This is the only of the presented measurement which has not been conducted on isolated particles in the gas phase.

6.1. ADAMANTANE-1-THIOL - THE WHOLE PICTURE

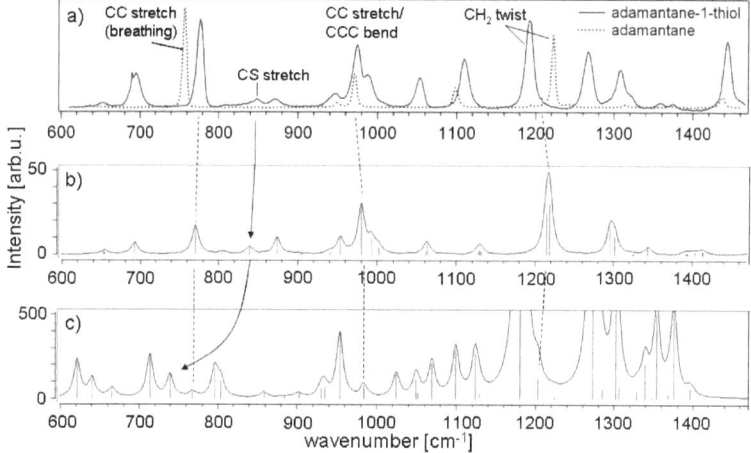

Figure 6.4: Experimental Raman spectrum for the adamantane-1-thiol (a) in comparison to the computed spectra for the neutral cluster (b) and the cation (c). In (a) the measured spectrum for adamantane is added for comparison (dotted line). The Raman peaks connected by arrows correspond to a vibrational mode of mainly SC stretch character. This mode has an energy of 92 meV in the cluster ion, in excellent agreement with the vibronic structure shown in Fig. 6.2. Note that the Raman intensities for the neutral cluster and the ion are on different scales.

of the other vibrational modes which do not directly involve vibrational motion of the thiol group. The modes that newly arise in the adamantane-1-thiol spectrum with respect to adamantane all exhibit the significant participation of the thiol group.

The XAS data for both thiolated (top) and pristine adamantane (bottom) are displayed in Fig. 6.5. X-ray absorption spectroscopy (XAS) data for adamantane and adamantane-1-thiol have been recorded at beamline 10.1 at the Stanford Synchrotron Radiation Lightsource, SLAC National Accelerator Laboratory[3] [119]. The absorption signal was measured while scanning the incident photon energy through the carbon K-edge and the resolution was determined to be 0.05 eV.

Upon thiol attachment the absorption onset as well as the two main resonances, which are found at 287.0 eV and 287.5 eV for adamantane, are only slightly shifted

---
[3]These data were kindly provided by Trevor Willey.

129

CHAPTER 6. RESULTS & DISCUSSION - PART II: MODIFIED DIAMONDOIDS

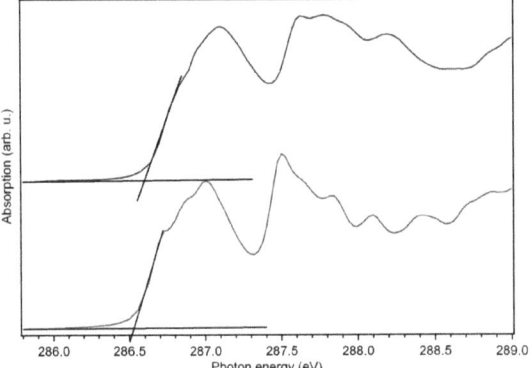

Figure 6.5: X-ray absorption data for adamantane-1-thiol (top) compared with pristine adamantane (bottom). The data reveal that the unoccupied states remain nearly unchanged upon thiolation with only a small shift of ~0.1 eV to higher energy. This suggests that the surface nature of the LUMO [24] remains relatively unchanged. The spectra have been scaled to same height to simplify comparison.

to higher energies by ~0.1 eV. In an earlier study these first two resonances could be linked to the CH- and the CH_2-environment of the diamondoid's surface atoms showing that the LUMO is determined by the hydrogen surface and is delocalized at the outside of the cluster [24]. The data for adamantane-1-thiol show that upon thiolation the unoccupied states remain largely unchanged in energy [24, 31, 42]. The electronic structure computations conducted within this thesis show that the LUMO also keeps its delocalized nature. Similar to adamantane the LUMO of adamantane-1-thiol in Fig. 6.6 is distributed around the cluster's hydrogen shell.

In Fig. 6.7 the optical spectra of the adamantane-1-thiol (top) and the pristine diamondoid (bottom) are compared. The absorption energies lie in the same regime. For the adamantane thiol a broad resonance is marking the absorption onset with an optical gap of 5.85 eV (as defined in section 5.1.3). This corresponds to a lowering of the optical gap of adamantane, which was measured to be 6.49 eV [90], by 0.64 eV due to thiolation. The peak of the first transition in the adamantane-1-thiol spectrum lies at 6.10 eV. The spectrum of adamantane exhibits numerous very sharp peaks which have been attributed to vibronic exci-

6.1. ADAMANTANE-1-THIOL - THE WHOLE PICTURE

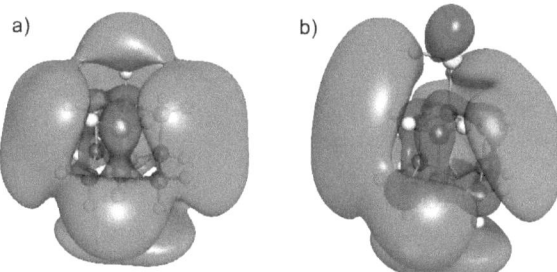

Figure 6.6: The lowest unoccupied molecular orbitals (LUMOs) of a) adamantane and b) adamantane-1-thiol calculated with sc Turbomole at B3LYP/6-31++G** level of theory. The delocalized nature of the LUMO of adamantane persists in the thiolated version.

tations and Rydberg states [32] while the thiol spectrum is mostly smooth with only a few, rather broad resonances. The lack of structure in the thiol optical spectrum also goes hand in hand with the disappearance of vibronic structure which is observed in Fig. 6.2 and is discussed above. The attachment of a thiol group leads to the lower C_s symmetry compared to T_d adamantane. Symmetry also implies that for the thiol, unlike adamantane, no restrictions due to dipole selection rules apply to the optical transitions. This is, next to the disappearance of Rydberg states, another reason for disappearance of sharp transitions. Symmetry constraints for the optical transitions in adamantane, however, do not concern the lowest transitions as seen in section 5.1.3 and are therefore not responsible for the differences in the optical gaps observed in Fig. 6.7 and listed in Tab. 6.1.

The same setup used to measure the luminescence of adamantane in section 5.2 was used to check for photoluminescence of adamantane-1-thiol. No photoluminescence in the energy range from 4.1 to 7.7 eV has been detected. The excitation energy was varied between 5.6 and 8.8 eV which covers band-to-band transitions from the optical gap to the ionization potential of adamantane-1-thiol (comp. Tab. 6.1). This means the UV luminescence found for adamantane [101] vanishes upon the attachment of a thiol group. To be below the detection limit it must be quenched by at least two orders of magnitude. This means that the ratio of radiative to nonradiative rate decreases by a factor of > 100.

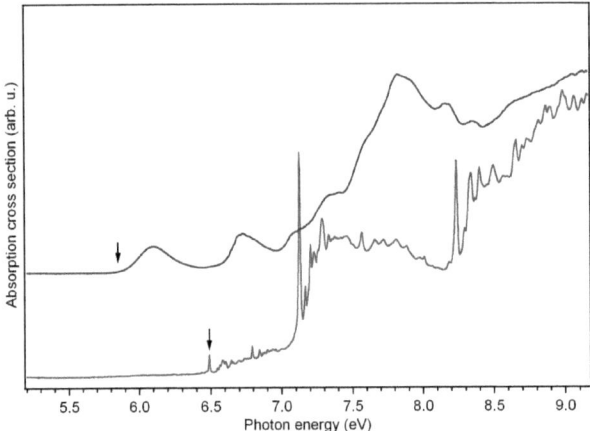

Figure 6.7: Optical absorption of adamantane-1-thiol (top) compared to that of pristine adamantane (bottom). Most obvious is the disappearance of the sharp resonances in the adamantane spectrum. Also the optical gap (arrows), as defined in Ref. [90], is lowered by 0.64 eV. The spectra have been scaled to similar height at their stronger absorbing bands beyond 7.5 eV.

In the previous paragraphs the experimental data are reported for the C1s core level, the highest occupied and lowest unoccupied electronic states and the optical transitions between them, and the Raman data. In the following the x-ray and ultraviolet photoelectron, x-ray absorption, Raman, and optical data are combined to give deeper insight into how the addition of a thiol group changes the properties of adamantane. Fig. 6.8 depicts a schematic graphical summary of the changes in the electronic structure.

Particularly poignant are the observed changes in the optical properties. Their discussion now benefits from the knowledge of the band edge data introduced above. In the following the changes in optical properties, i.e., band-to-band transitions, are compared to changes in the band edges, i.e., changes in an experimentally constructed HOMO-LUMO gap[4], which are reflected by the band edge measurements presented at the beginning of this section.

Considering the spectra in Figs. 6.2 and 6.5 the changes in the optical response

---

[4]Note that this is not a HOMO-LUMO gap in the strict sense due to the experimental effects discussed in section 5.1.3.

6.1. ADAMANTANE-1-THIOL - THE WHOLE PICTURE

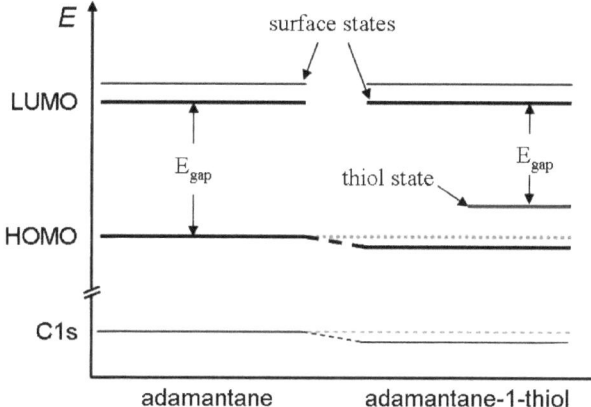

Figure 6.8: Schematic drawing of the electronic levels of adamantane and adamantane-1-thiol. In the functionalized diamondoid the S electron lone-pair gives rise to a "thiol state" constituting the HOMO of adamantane-1-thiol and, effectively, lowering the gap by ~ 0.6 eV.

primarily arise from changes in the occupied states. The red-shift of 0.64 eV of the optical gap can be accounted for by the additional electron lone-pair state that arises for the thiol 0.58 eV above the band edge of adamantane (Fig. 6.2). This lowering of the optical gap leads to a gap value for adamantane-1-thiol that is comparable to the values of penta- and hexamantanes reported in chapter 5. However, the optical gap is of a different nature because the transition takes place between the surface dominated LUMO and the HOMO which is no longer localized inside the cluster, as is the case for pristine diamondoids. Instead, the HOMO is now given by the electron lone-pair state of the thiol group as apparent from Fig. 6.3. The first two noticeable spectral features in the optical absorption, shown in Fig. 6.7, are centered around 6.1 eV and 6.7 eV with an energy spacing of 0.6 eV. This energy spacing is in good agreement with the first two manifolds of resonances in the XAS spectrum in Fig. 6.5. The broadness of these two peaks can be explained with the convolution of the thiol state and the respective features in the XAS spectrum shown in Fig. 6.2. Considering the C_s symmetry of adamantane-1-thiol it can be safely concluded that the first two resonances in the optical spectrum in Fig. 6.7 are transitions from the HOMO because there are no dipole-forbidden transitions in structures of this particular molecular point group. A "main resonance" can be located around 7.8 eV for the thiol. This resonance appears in the spectrum of the pristine diamondoid somewhat disguised by Rydberg resonances and shifted to lower energies by some $300-400$ meV. This corresponds to the shift of more than 0.2 eV which has been observed in the photoelectron data (Fig. 6.2) and the additional small shift in the unoccupied states (Fig. 6.5). The discussion above shows that most of the changes that occur in the optical response and the optical gap can be reconstructed from comparison with occupied and unoccupied states and the HOMO-LUMO gap, respectively.

As will be seen in the following section which discusses optical transitions in larger diamondoid thiols there are indications that a very weak optical transition occurs at energies around 5.5 eV. This dissociative excitation is too weak, however, to be identified from the present adamantane thiol data alone.

The lack of detectable photoluminescence for adamantane-1-thiol suggests that UV photoluminescence found in adamantane is quenched by the attachment of a thiol group. It was shown that the electron lone-pair of the thiol group gives rise to an additional density of states above the adamantane HOMO. Because luminescence typically occurs due to transitions from the LUMO to the HOMO the altered nature of the HOMO is to be held accountable for the disappearance of the luminescence.

In the interpretation of the results for the valence states the shoulder of the thiol state in the photoelectron spectrum in Fig. 6.2 was tentatively assigned to vibrational progressions. A comparison with the supporting Raman data helps to underpin and specify the assignment made. The five gaussian peaks used to fit the thiol state in Fig. 6.2 have an almost equidistant energy spacing of approximately 90 meV. The Raman peak in the spectrum computed for the ionic cluster that comes the closest is a mode of medium intensity at 739 cm^{-1} (92 meV). The computations show that this Raman signal belongs to an SC stretch mode. The same SC stretch mode is found in the neutral cluster at considerably higher wavenumbers, dropping from 839 cm^{-1} in the neutral cluster to 739 cm^{-1} in the cluster ion (follow the arrow in Fig. 6.4). This drastic change indicates that the SC stretch mode is strongly excited upon ionization. Next to the match in energy this is additional evidence that the SC stretch mode is responsible for the shoulder of the thiol peak in the photoelectron spectrum of adamantane-1-thiol.

As seen above, the LUMO of adamantane, unlike the HOMO, remains almost completely unaffected by the surface functionalization. This applies to the energy of the electronic states as well as to their delocalization at the nanodiamond's

6.2. Larger Diamondoid Thiols

	adamantane-1-thiol ($C_{10}H_{15}$-SH)	adamantane ($C_{10}H_{16}$)	Δ
optical gap	5.85 ± 0.06	6.49 ± 0.03 [90]	-0.64
adiabatic IP	8.65 ± 0.04	9.23 ± 0.11 [25]	-0.58
cond. band edge	286.6 ± 0.1	286.5 ± 0.1 [24]	0.1
core level	289.9 ± 0.1	289.7 ± 0.1	0.2

Table 6.1: Electronic key figures of adamantane-1-thiol compared to literature values for adamantane. Δ gives the difference between values for adamantane-1-thiol and adamantane. Negative values correspond to a decrease of the corresponding values upon thiolation, positive values to an increase. All values are given in eV.

surface. The present data therefore suggest that properties of diamondoids that are intimately linked to the particular nature of their LUMO will withstand thiolation. This is the case for the technologically relevant negative electron affinity (NEA) which has been predicted for pristine diamondoids [31] and has recently been verified experimentally for SAMs of [121]tetramantane-6-thiols [2, 46]. However, properties involving the LUMO and any additional electronic levels (e.g., the HOMO) may not persist, as seen in the case of photoluminescence.

6.2 Larger Diamondoid Thiols

Investigating the effect of a thiol group on the electronic structure of the smallest diamondoid, adamantane, revealed drastic changes in the optical properties. In this section changes in the optical properties of diamondoid thiols are investigated as a function of diamondoid size and of functionalization site. For this purpose, several thiolated diamondoids ranging in size up to tetramantane have been investigated using optical absorption spectroscopy. As seen above in section 6.1, the changes in the optical properties are caused by changes in the occupied states. Therefore, additional photoelectron spectroscopy measurements have been performed on the diamondoid thiols to help explain the observed changes. Also, in a collaboration with the group of Prof. Bonačić-Koutecký from the Humboldt University Berlin, the optical spectra of diamondoid thiols were computed to support the interpretation of the spectra. The computations presented in this section have been performed by the Bonačić-Koutecký group. The combination

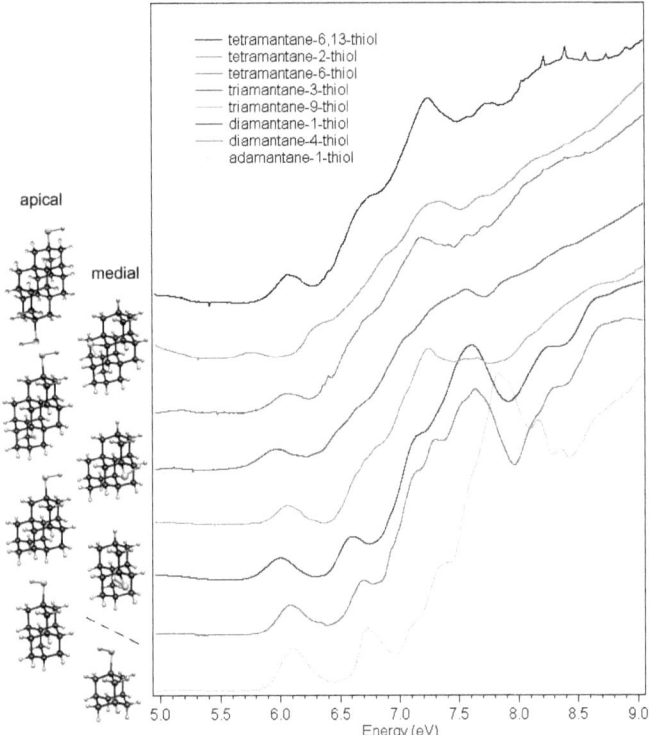

Figure 6.9: The optical absorption spectra of the measured thiolated diamondoids. The structure of each diamondoid is shown to the left of the corresponding curve. For diamantane through tetramantane isomers with apical and medial thiol groups are investigated.

of experimental and theoretical work greatly improved the understanding of the optical properties of diamondoid thiols which lead to a joint publication of the results in *The Journal of Chemical Physics*.[5] The computational results of the publication are presented in this section (6.2) not without explicitly pointing out that, in this one particular case, the calculations were *not* performed by the author of this thesis.

---

[5]L. Landt et al., *Experimental and theoretical study of the absorption properties of thiolated diamondoids*, The Journal of Chemical Physics **132**, 144305 (2010) - Ref. [120]

6.2. LARGER DIAMONDOID THIOLS

In general, thiol functionalized diamondoids differ both by the topology of the carbon framework as well as by the substitution site of the thiol group. In this work only diamondoids are considered where the thiol group is attached to a tertiary carbon replacing the hydrogen of a CH group. The functionalized diamondoids range in size from adamantane to tetramantane. In the latter case only the rod-shaped 1D isomer, [121] tetramantane, is considered. The substitution of the hydrogen atom at a tertiary CH group by a thiol group leads for the investigated species in principle to 14 different types of functionalized diamondoids [121, 122]. Additionally to singly thiolated diamondoids, a diamondoid with two equivalent thiol groups, [121] tetramantane-6,13-dithiol, is investigated. The absorption spectra of the investigated diamondoid thiols are presented in Fig. 6.9 and the structures are shown next to the spectra. Their systematic nomenclature is given in the legend of the graph.

For all measured diamondoid thiols the absorption spectra in Fig. 6.9 are fairly smooth and featureless resembling in their general appearance the spectrum of adamantane-1-thiol discussed in the previous section. Qualitatively, all spectra exhibit weak bands in the low energy region up to approximately 7.0 eV and a broad, strong absorption extending from 7.0 eV to 9.0 eV. The strong changes in the spectral characteristics that were observed for pristine diamondoids in chapter 5 do not persist for diamondoid thiols. Instead, a broad peak at approximately 6 eV marks the first strong transition for all diamondoid thiols. A size-dependent shift of this first major transition, however, is not apparent from Fig. 6.9. Also, very faintly, a signal can be made out between 5.0 and 5.5 eV for some diamondoids. First, the obvious signal at 6 eV is investigated. The small feature will be scrutinized with the help of theoretical investigations later on.[6]

Fig. 6.10 shows an enlarged view of the absorption onset of the spectra. Further, the spectra are sorted by the functionalization site of the diamondoid thiol. The spectra of diamondoid thiols with a thiol group at the *apical* position (i.e., at the tip of diamondoid, compare Fig. 6.9) are stacked in the left panel. Those with *medial*, i.e., sidewise attached thiol groups, are shown in the right panel. It is apparent from Fig. 6.10 that for apical isomers the energetic position of the peak at 6.1 eV is entirely independent of the diamondoid size. For adamantane-1-thiol

---

[6]Note that the sharp little dips around 5.5 eV which are present in some of the spectra are due to functioning errors of the Keithley picoampèremeter during the measurements and the spikes between 8.5 and 9 eV in the tetramantane-dithiol are due to outgassing of o-rings at high temperatures.

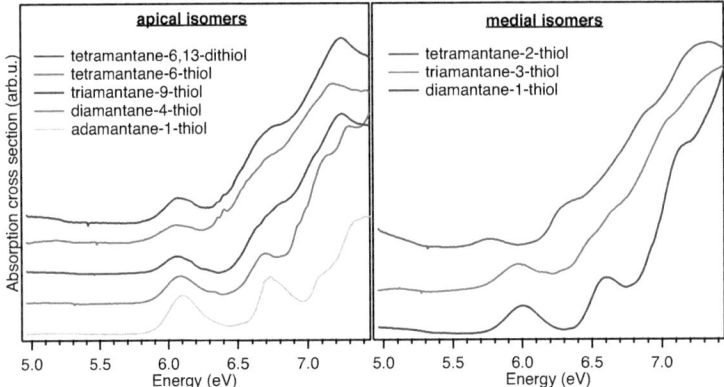

Figure 6.10: The absorption spectra of diamondoid thiols grouped by their functionalization site: a) Diamondoids with an apical thiol group and b) diamondoids with a medial thiol group.

this peak has been attributed in the previous section to transitions from the thiol-dominated HOMO to the diamondoid-like LUMO. The relative intensity of the peak diminishes with size in accordance decreasing atomic proportion of a thiol group in a growing diamondoid thiol.

This observation also remains true for the doubly functionalized [121] tetramantane-6,13-dithiol which exhibits a peak at the same energy with approximately twice the relative intensity. A second broad resonance which is observed for adamantane-1-thiol at 6.8 eV is in larger diamondoids disguised by a steep increase in signal intensity. At sizes of triamantane and beyond it is only visible as a shoulder on the steep spectral flank. Another interesting observation that can be made from Fig. 6.10 is that for tetramantane-6-thiol a sharp feature reappears at 6.4 eV which is characteristic for the pristine cluster (compare Fig. 5.3). Resembling the case of [121] tetramantane, the spectral feature of the thiol consists of a double peak with a spacing of ∼50 meV. It appears shifted by 100 meV towards higher energies. This reappearance of spectral features can be interpreted as a first faint sign that the dominant influence in the optical properties of the thiol group over the diamondoid begins to wane with increasing diamondoid size.

Comparison to the investigated medial isomers of diamantane and triamantane in the right panel shows that the first intense peak lies shifted by more than 100 meV at approximately 6.0 eV. In section 6.1 it has been shown that the changes in

6.2. LARGER DIAMONDOID THIOLS

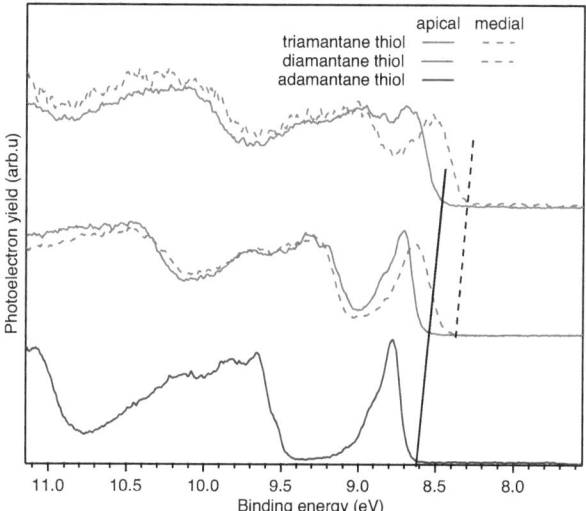

Figure 6.11: The photoelectron spectra of some thiolated lower diamondoids. Solid lines represent diamondoids with an apical thiol group, dashed spectra belong to diamondoids with medial thiol groups.

the optical properties upon thiolation, and in particular this peak, originate in changes in the valence states. To investigate the role of the valence states for the optical properties of larger diamondoid thiols photoelectron spectroscopy is employed. Fig. 6.11 compares the photoelectron spectra of diamondoids with apical and with medial thiol groups. The spectra of apical and medial isomers are shown as solid and dashed lines, respectively. The high energy parts of the spectra (beyond $\sim 9\,\mathrm{eV}$) superimpose on the spectra of the pristine diamondoid. The valence band edges of the medial isomers are slightly shifted towards lower binding energies compared to the apical isomers. As in the case of adamantane in section 6.1, the effect is due to a high electron affinity of the thiol group which leads to a redistribution of the electron density within the cluster. Electron density is withdrawn from the diamondoid part which increases the binding energy of the remaining electrons [5]. The effect is larger for medially attached thiol groups due to their proximity to large parts of the cluster. In turn, the accumulation of electron density leads to lower binding energies for electrons belonging to the thiol group.

Comparison of Figs. 6.10 and 6.11 shows that the little and yet noticeable size dependence of the valence band is not reflected in the optical spectra. This is true despite the fact, that diamondoid thiols do not exhibit a size dependence of the LUMO as seen in section 6.1 for adamantane thiol and confirmed in experiments on larger diamondoid thiols [123]. A possible explanation is provided by the size dependent electron-hole interaction in the excited state. As already seen in section 5.1.3 such an interaction has a size-dependence that counteracts the shift in the valence band. The comparable magnitude of the effect, however, is purely accidental.

A closer look at the optical spectra of the medial isomers in Fig. 6.10 reveals that for tetramantane the discussed peak appears already at approximately 5.8 eV. A more detailed structural analysis, however, shows that the medial isomers of dia- and triamantane differ from that of tetramantane. Medial thiol groups can be bound to a diamondoid via a carbon atom that is either connected to two tertiary carbons (CH groups) and a secondary carbon (a CH_2 group) or it is surrounded by two tertiary and a quarternary carbon. In the following, to distinguish the different chemical environments the former are labeled *medial1*-isomers and the latter *medial2*-isomers. Diamantane thiol is the smallest species that gives rise to the *medial1*-thiols, while *medial2*-isomers are present for thiolated triamantane and higher diamondoids. One important effect leading to differences in the electronic levels is the redistribution of electron density discussed above.

Another way in which the three isomers differ is by their steric interaction. For the *medial1*-isomers the steric interaction consist of one pair of 1,3-diaxial interactions with respect to SH, while there are two pairs of 1,3-diaxial interactions for *medial2*-isomers. The *apical*-isomers do not exhibit 1,3-diaxial interactions. The nature of the 1,3-diaxial steric interaction is visualized in Fig. 6.12 using the example of cyclohexane. The thiol group is interacting with the hydrogen atoms which are bound along the same axis.

Complementary theoretical investigations are carried out to explore the effects that different functionalization sites have on the optical absorption of diamondoid thiols. The absorption spectra were computed by the Bonačić-Koutecký group using the coupled cluster method (CC2) as implemented in TURBOMOLE. Prior, geometric structures of the diamondoid thiols were optimized at B3LYP/6-311++G** level of theory.

In Fig. 6.13, the experimental and computed absorption spectra are compared. The theoretical and the experimental transition energies are in good agreement and the deviation in all studied cases is smaller than 0.2 eV. The theoretical spec-

6.2. LARGER DIAMONDOID THIOLS

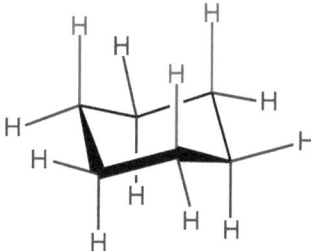

Figure 6.12: Schematic drawing of the steric 1,3-diaxial interaction in cyclohexane. A set of stericly interacting hydrogens, e.g., are extending straight up from the carbon ring.

tra reveal a very weak band centered below 5.5 eV. This band is also observable in some of the experimental spectra in Figs. 6.9 and 6.10. An experimental determination of the optical gaps as defined in section 5.1.3 is impractical in the case of diamondoid thiols. This is due to the very broad resonances of the experimental spectra which are low in intensity. Also, the weak bands below 5.5 eV in some cases extend beyond the recorded energy region as can be seen from Fig. 6.13 a).

A more intense band is observed between 5.5 and 6 eV. The calculated oscillator strengths in this energy range are very low and therefore, only qualitative comparisons of experimental and theoretical intensities are possible. At energies higher than 6.3 eV, the intense absorption bands are well reproduced by theory, exhibiting a shift to lower energies with increasing diamondoid size as shown in Fig. 6.13. Furthermore, the transition energy of the second absorption band is specific for the substitution-site and differs among the types of isomers as already seen in Fig. 6.10. Similar to the experimental data, theory finds a red shift of 0.1 - 0.2 eV for *medial1*-isomers and a shift of 0.4 - 0.5 eV for *medial2*-isomers with respect to *apical*-isomers. Also, the location of the band is only slightly dependent on the diamondoid size.

In order to assign the transitions to the diamondoid and to the thiol fragments the character of the excited states of the thiol functionalized diamondoids in terms of leading configurations contributing to the transitions is investigated. For this characterization it is useful to compare characteristic features of low lying excited states with those of alkane thiols. The calculations show that the first excited state for all investigated species corresponds to the n-σ^*_{SH} transition that is also

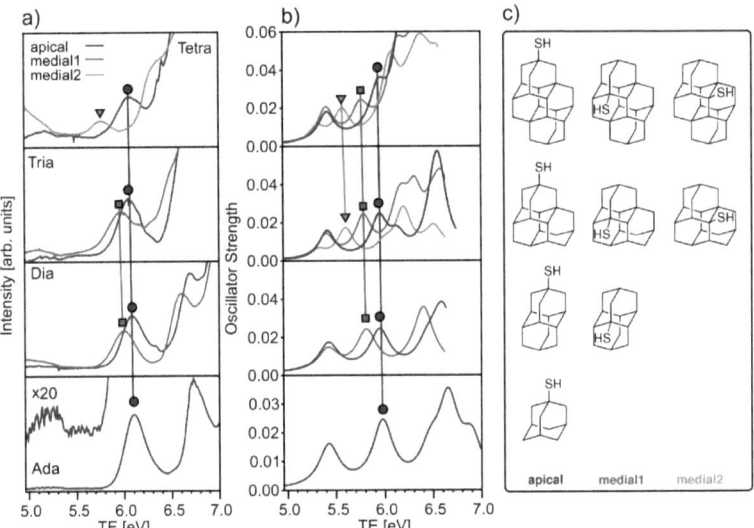

Figure 6.13: Comparison of a) experimental and b) CC2 computed spectra of diamondoid thiols. The peak markers are of different shape for *apical*- (●), *medial1*- (■) and *medial2*-isomers (▼). The vertical lines connect the band maxima of the the n-σ_{SC}^* transitions. The structures are presented in panel c).

present in alkane thiols [124, 125, 126]. The lowest excited states in methane thiol correspond to an excitation from a nonbonding p-orbital on sulfur (n) to the σ^* orbitals, which have antibonding character along the sulfur-hydrogen (SH) and sulfur-carbon (SC) bonds, respectively. Although partial Rydberg character is present, especially for the second excited state, it is convenient to simplify the notation according to n-σ_{SH}^* for the transition to the first excited state and according to n-σ_{SC}^* for the transition to the second excited state of the thiols. The importance of the σ_{SC}^* character for the second excited state of methane thiol was also found in photodissociation experiments [126]. The comparison of the absorption spectra of adamantane thiol and methane thiol demonstrates the analogous character of the first two excited states that can be also seen from the density differences between these excited states and the ground state as presented in Figure 6.14. From this similarity in density differences a direct correspondence of the first two excited states can be derived. These transitions consist of several excitations, all involving the HOMO (being dominantly the nonbonding p-orbital

6.2. LARGER DIAMONDOID THIOLS

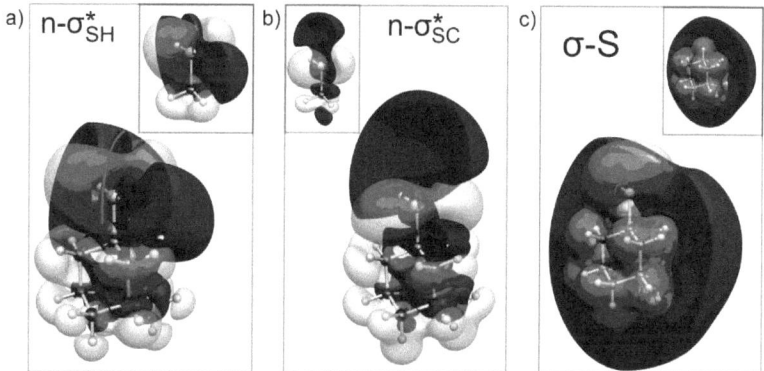

Figure 6.14: Electron density difference between excited states and the ground state of adamantane thiol corresponding to a) n-σ^*_{SH}, b) n-σ^*_{SC} and c) σ-S transitions. Accumulation of density is depicted in dark, depletion of density in light grey. A line in a) indicates the nodal plane in the σ^*_{SH}. The inserts show the density difference for the corresponding excited state of the molecular subunits methane thiol (a and b) and adamantane (c).

of sulfur as seen in section 6.1), into a large number of highly delocalized virtual orbitals, which are necessary to generate a localized excitation. The first and second excited states of all diamantane thiols, triamantane thiols as well as of the *apical-* and *medial1*-[121]tetramantane thiols are characterized by these n-σ^*_{SH} and n-σ^*_{SC} type of transitions, respectively. Only in the case of *medial2*-[121]tetramantane thiol the n-σ^*_{SC} transition corresponds to the third excited state.

The n-σ^*_{SC} transition energies obtained experimentally and theoretically are isomer specific. Correlations with the molecular properties of the diamondoid thiols, as presented in Figure 6.15 help to qualitatively explain this behavior. The energies of the n-σ^*_{SC} transitions (Figs. 6.15 a and 6.15 b) correlate well with the computed energies of the HOMO (Fig. 6.15 c), which are higher for the *medial1*-isomers and *medial2*-isomers with respect to those of *apical*-isomers in agreement with the experimental data in Fig. 6.11. Furthermore, the calculations find a correlation between the SC bond lengths and the n-σ^*_{SC} transition energy. As depicted in Fig. 6.15 d the SC bond lengths are slightly elongated for the *medial1*-isomers and more elongated for *medial2*-isomers with respect to *apical*-isomers, indicating weaker SC bonds. These weaker bonds cause a red shift in n-σ^*_{SC}

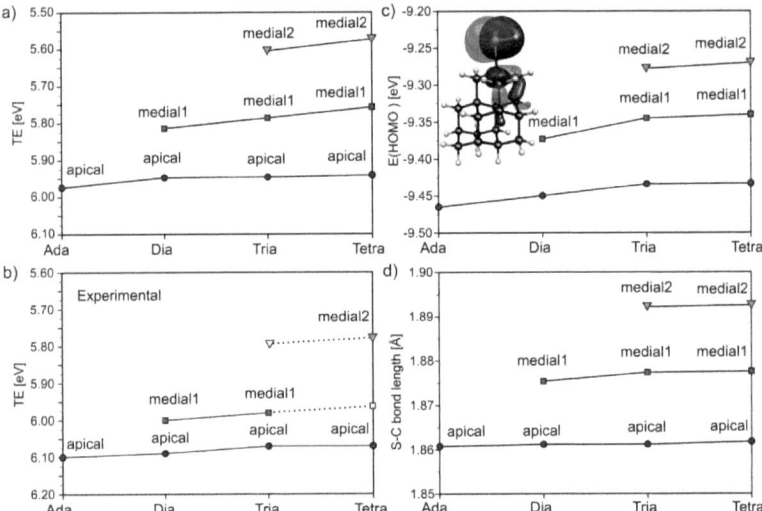

Figure 6.15: a) Theoretical and b) experimental n-σ_{SC}^* transition energies (TE) for *apical* (●), *medial1* (■) and *medial2* (▼) classes of isomers with increasing diamondoid size from adamantane (Ada), diamantane (Dia), triamantane (Tria) to tetramantane (Tetra). c) HOMO energies and d) SC bond length for *apical-* (●), *medial1-* (■) and *medial2-*isomers (▼) with increasing diamondoid size. The electron density distribution of the HOMO of apical-triamantane thiol is shown in the insert of c). Good agreement of theoretical values with the experimental data is obtained for the behavior of the n-σ_{SC}^* transition energies. This allows for extrapolation of the experimental data labeled by dotted lines. Panel c) confirms that different transition energies arise mainly from differences in the occupied states. Lower transition energies go hand in hand with longer bond lengths, as shown in panel d).

6.2. LARGER DIAMONDOID THIOLS

transition energies for *medial1*- and *medial2*-isomers. Both correlations can be interpreted as a result of the influence of the immediate chemical environment on the C-SH group. This influence consists of zero (*apical*), one (*medial1*) or two (*medial2*) pairs of steric 1,3-diaxial interactions. The Coulomb-type interaction with the sulfur lone pairs, which dominate the HOMO (compare insert of Fig. 6.15c), raises the HOMO's energy isomer specifically while the elongation of the SC bond is due to steric repulsion. The interplay of both effects is therefore responsible for the strong isomer specific n-σ_{SC}^* energy, while the n-σ_{SH}^* transition energy is less sensitive to the type of isomer supporting our quantitative experimental and theoretical findings.

The UV photoluminescence of diamondoids which has been found in chapter 5 is a feature of possible technological interest. As mentioned, thiolation provides viable means to incorporate diamondoids into photonic devices. Just like for adamantane-1-thiol, however, no photoluminescence could be detected for the larger diamondoid thiols investigated in this work. The quenching of the UV photoluminescence of diamondoids is likely to be linked to low energy excited states in diamondoid thiols. Better understanding of the underlying mechanisms is required.

To identify the precise causes for the lack of photoluminescence, the possible excited state decay mechanisms of the low-energy excited states of diamondoid thiols have been investigated theoretically. Geometry optimizations in the first excited state (n-σ_{SH}^*) of adamantane thiol, performed in the group of Prof. Bonačić-Koutecký, resulted in SH bond breaking [120], therefore this state cannot lead to fluorescence. The theoretical results allow to identify the σ-S Rydberg transition in diamondoid thiols that in unfunctionalized diamondoids is typically responsible for their fluorescence. In adamantane the σ-S transition to a Rydberg state [127, 32] has been found at 6.9 eV with the main contribution from the HOMO−1 → LUMO excitation. As expected from the measurements and their interpretation in section 6.1, the HOMO−1 of larger diamondoid thiols has a σ character with respect to the diamondoid framework and the LUMO has a very diffuse s-type character that is specific also for higher diamondoids, as already measured and theoretically predicted [42, 128]. This transition leads to the electron density redistribution given in Fig. 6.14 c that allows the assignment of the purely diamondoid-like σ-S transition to the 6th excited state for adamantane thiol. Based on this analysis the lowest energy diamondoid Rydberg S states could also be identified for all other diamondoid thiols. Since the geometry opti-

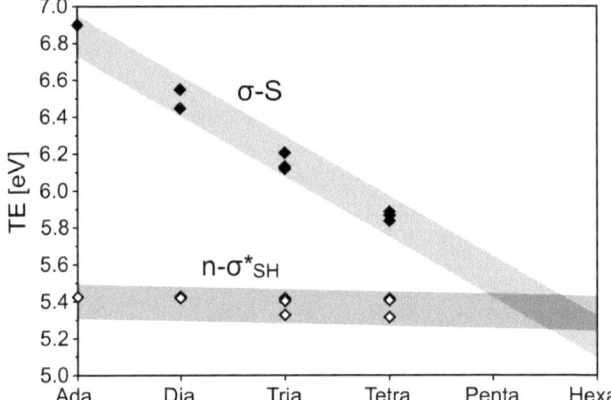

Figure 6.16: Size dependence of σ-S (◆) and n-σ^*_{SH} (◊) transition energies. The n-σ^*_{SH} (◊) transition, which is the energetically lowest transition for all small diamondoid thiols, exhibits almost no size dependence. Contrary, the σ-S (◆) transition, which is directly linked to the diamondoid part of the system, shows a strong dependence on the size of the diamondoid. Extrapolation of the present data, indicated by grey bars, predict a crossover at diamondoid sizes of five to six adamantane cage units. Beyond this point it is to be expected that diamondoid-like transitions will be the energetically lowest transitions in diamondoid thiols.

mization of adamantane thiol in this excited state gives rise to a bound state, it is expected that this state remains bound also for the higher diamondoid thiols. To explore the possibility of tuning the fluorescence properties in thiol functionalized diamondoids, which is desirable for the development of new optical materials, the size dependence of the position of the purely diamondoid states is investigated. In Fig. 6.16 the energies of the first purely diamondoid-like, σ-S-type state are shown together with the energies of n-σ^*_{SH} states as a function of diamondoid size. The dissociative σ^*_{SH} state is likely to be responsible for the quenching of luminescence in diamondoid thiols. Species in which S-type states lie energetically below these states, however, are good candidates for exhibiting fluorescence. Fig. 6.16 demonstrates qualitatively why fluorescence from diamondoid thiols smaller than pentamantane is not observed. By increasing the size of the diamondoid subunit, however, the purely diamondoid-like σ-S transition shifts to the red in thiol functionalized diamondoids (Fig. 6.16). Therefore, the results suggest that by increasing the size of the diamondoid subunit, fluorescent

6.2. LARGER DIAMONDOID THIOLS

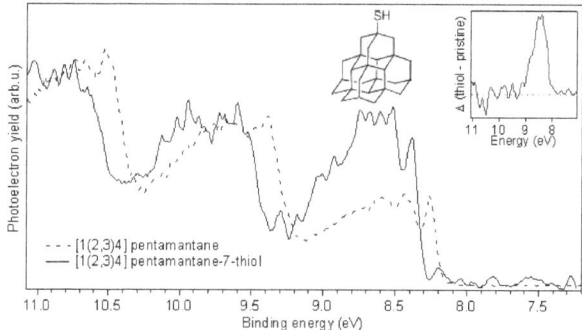

Figure 6.17: Comparison of the photoelectron spectra of [1(2,3)4] pentamantane and [1(2,3)4] pentamantane-7-thiol (structural inset). The spectrum of the thiol is shifted to higher binding energies by 160 meV. The spectral inset shows the intensity difference of the thiol and the diamondoid spectrum (when the shift is accounted for). The subtraction of the two spectra is able to retrieve the thiol state which is located around 8.5 eV.

species may be obtained for systems larger than pentamantane thiol.

It was not possible to experimentally check for the photoluminescence of larger diamondoid thiols so far. Severe experimental difficulties arise from the very low sample quantities available for diamondoid thiols of this size and, more decisively, from the temperature limitations of the gas-cell setup. Ever larger and heavier compounds require higher temperatures to reach a vapor pressure that is sufficient for the current absorption / photoluminescence measurements.

To check the plausibility of the predictions above photoelectron spectroscopy, which has a higher sensitivity due to the detection of electrons rather than photons, has been performed on a thiolated pentamantane. The photoelectron spectrum of [1(2,3)4] pentamantane-7-thiol, mapping the occupied states, has been recorded. In Fig. 6.17 the spectrum is compared to that of the pristine diamondoid. As observed previously for the diamondoid-like part, the spectrum is shifted to higher binding energies. This has been explained in section 6.1 with the electronegativity of the thiol group withdrawing electron density from the main body of the cluster. The spectra in Fig. 6.17 have been scaled to the same height for better comparability. It is apparent that the relative intensities of the first three broad resonances in the two spectra do not match. To shed light on the whereabouts of the thiol contribution the photoelectron signal of pristine diamondoid has been shifted to overlay the thiol spectrum and then subtracted from it. The spectral inset showing the residual thiol signal reveals the position of the extra density of states due to the non-bonding electron lone pair of the thiol. The comparison shows that for diamondoids larger than tetramantane transitions between diamondoid-like electronic states should already occur at energies below the n-σ_{SC}^* excitation. As seen above, the lowest energy excitation is the n-σ_{SH}^* excitation which is likely responsible for the quenching of the luminescence. This transition lies approximately 0.5-0.8 eV lower than the n-σ_{SC}^* excitation. Further increase of the diamondoid size, i.e., an additional shift of the diamondoid-like valence states, should lead to the predicted gas-phase photoluminescence for larger diamondoid thiols. For diamondoid thiols bound to surfaces, as in the case of diamondoid SAMs [2, 59], larger size diamondoids might not even be necessary. The functionalized diamondoids are likely to bind to the noble metal surface in the form of thiolates [129], i.e., if they split off the hydrogen atom at the end of the thiol group in favor of a stronger S-Au bond [130], n-σ_{SH}^* no longer occur. The lowest optical excitations in this case would be either the n-σ_{SC}^* excitation or the diamondoid-like σ-S-type transition. The latter prevails already in diamondoid thiols larger than tetramantane as demonstrated above.

The above investigations show that in diamondoid thiols the optical properties at the absorption onset are dominated by the thiol group. Two major transitions involving the thiol group have been identified in the diamondoid thiols: The dissociative n-σ_{SH}^* excitation and the stronger n-σ_{SC}^* excitation. These are the two lowest energy excitations in the experimentally investigated diamondoids. Neither of these transitions exhibits a notable dependence on the size of the diamondoid. The first excitation which is of purely diamondoid-like nature is of σ-S-type and is strongly size-dependent. Therefore it will for larger diamondoids surpass first the n-σ_{SC}^* and then the n-σ_{SH}^* excitation. This opens new opportunities for photonic devices based on diamondoid thiols.

6.3 Bulk-substituted Diamondoids

In this section the effect of the incorporation of impurity atoms into the diamondoid carbon framework is investigated. In particular, the influence of oxygen

6.3. BULK-SUBSTITUTED DIAMONDOIDS

substituents on the optical properties is examined. For this purpose the absorption spectra of four bulk-substituted diamondoid species are recorded, which are shown in Fig. 6.18: three oxadiamondoids (b)-(d) are investigated to determine the influence of oxygen incorporations on the optical properties of diamondoids. Additionally, urotropine (a), an adamantane cage with four nitrogen substituents, is investigated to learn more about the influence of nitrogen in the diamondoid lattice. Typically, the two elements play very different roles in diamond.

Above all, these investigations provide information about how impurity atoms affect the optical properties of diamondoids. The precise structural control in diamondoids, however, also makes these investigations interesting models for oxygen and nitrogen substituents in a local diamond environment.

Note that the replacement of a carbon atom does not result in "doping" of the diamondoid because no extra electrons are provided to the system. Instead the additional valence of both nitrogen and oxygen substituents is accounted for by the loss of one or two surface hydrogens, respectively. Although the substitution of atoms on regular bulk diamond lattice sites by single oxygen or nitrogen atoms will not function as doping, i.e., add free electrons or holes to the host crystal, it is a first step to understand the influence of impurity atoms on the optical properties of nanoscale diamond.

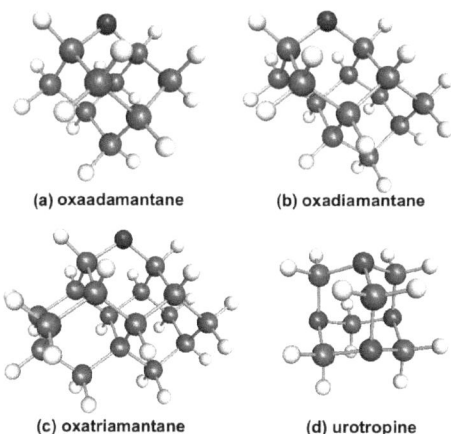

(a) oxaadamantane (b) oxadiamantane

(c) oxatriamantane (d) urotropine

Figure 6.18: Ball and stick models of the investigated bulk-substituted diamondoids: (a) oxaadamantane C_9OH_{14}, (b) oxadiamantane $C_{13}OH_{18}$, (c) oxatriamantane $C_{17}OH_{22}$ and (d) urotropine $C_6N_4H_{12}$.

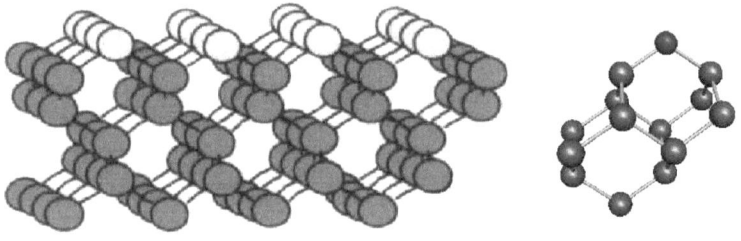

(a) oxygenated diamond (100) surface   (b) oxadiamondoid

Figure 6.19: (a) The structure of an oxygenated (100) diamond surface in the ether (C-O-C) configuration [131]. (b) The structure of oxadiamantane (hydrogen omitted).

Oxygen

Oxygen occurs mainly on diamond surfaces and plays a significant role in determining their properties [131]. In the investigated oxadiamondoids a secondary carbon atom is replaced by an oxygen atom.[7] This constellation is typical of an ether bridge (C-O-C) oxygen passivation of the technologically important diamond (100) surfaces [132], which is shown in Fig. 6.19. Oxadiamondoids therefore provide a nanoscale model to study single oxygen surface atoms in a local diamond lattice framework.

To study the influence of oxygen-substitution on the optical properties the absorption spectra for the three oxadiamondoids in Fig. 6.18 has been recorded in the gas phase. The oxadiamondoid spectra are shown in Fig. 6.20 and are compared to the spectra of the corresponding pristine diamondoids (dashed). The spectra of the oxadiamondoids have some general characteristics in common: Two very sharp peaks at the absorption onset are followed by further, less intense features. The rest of the spectrum is comparably smooth and after a shallow local minimum that lies approximately 0.5 eV higher in energy the absorption continuously rises. Only for oxadiamantane a second set of pronounced spectral features are observed 1 eV higher in energy than the features at the onset. Oxatriamantane has only a few very small, irregular features at higher energies and for oxaadamantane the spectrum remains completely smooth at energies beyond the first features. Curiously, the extra series of features for diamantane, which is shifted by 1.06 eV with respect to the first appearance, exhibits a very similar energy spacing to the

---

[7] Due to the lower valence of oxygen, effectively, the entire CH_2 group is replaced.

6.3. BULK-SUBSTITUTED DIAMONDOIDS

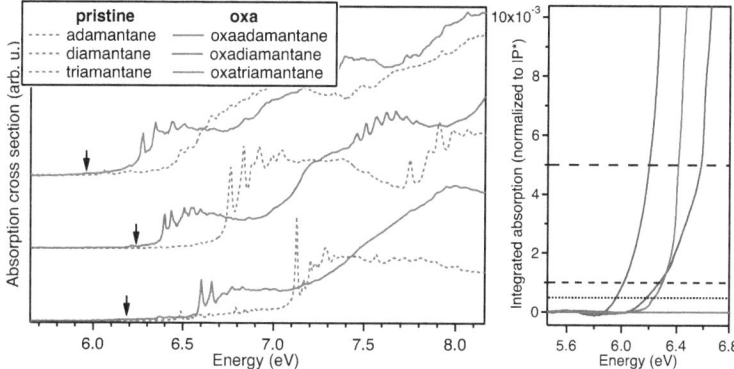

Figure 6.20: Left: The optical absorption spectra of oxadiamondoids and their pristine counterparts (dashed). The optical gaps ($5 \cdot 10^{-4}$) are indicated by arrows. Right: Integrated absorption signal of oxadiamondoids. Because the ionization potentials of oxadiamondoids have not yet been determined the integrated absorption has been normalized to the ionization potentials of the corresponding pristine species [26]. The dashed, finely dashed, and dotted lines indicate thresholds of $5 \cdot 10^{-3}$, $1 \cdot 10^{-3}$, and $5 \cdot 10^{-4}$, respectively.

first five peaks. It appears that the sharp regular features at the absorption onset repeat themselves at higher energies.

A second observation is that the spectra of oxadiamondoids shift towards lower energies with increasing diamondoid size. This behavior is similar to the pristine species and as one would expect from the quantum confinement model but contrary to the observation made for diamondoid thiols.

The positions of the optical gaps are indicated by arrows in Fig. 6.20. They are determined from the integrated absorption signal shown in the right panel of Fig. 6.20 according to the method introduced in section 5.1.3. However, the ionization potentials of the pristine species have been used to normalize the integrated signals because the ionization potentials of the oxadiamondoids have not yet been determined. The resulting values for a threshold of $5 \cdot 10^{-4}$ are 6.18, 6.24 and 5.96 eV for oxaada-, oxadia-, and oxatriamantane, respectively, with an error on the order of 0.1 eV. This compares to 6.49, 6.40, and 6.06 eV for the corresponding pristine species. The difference in the optical gap between the pristine and the oxa-species decreases with increasing diamondoid size from 300 meV for adamantane to 100 meV for triamantane. Interestingly, the deviation of the optical gap from a linear trend which is observed for diamantane is also present in oxadiamondoids.

In pristine diamondoids this could be explained by the fact that the HOMO-LUMO transition is dipole-forbidden due to the symmetry of diamantane. However, Oxadiamantane has considerably lower C_s symmetry. Within this point group no forbidden transitions exist (compare Appendix) and thus the selection rules cannot explain the observed deviation. Yet, the gap of oxadiamantane is even larger than that of oxaadamantane. For oxaadamantane and oxatriamantane, which both exhibit C_{2v} symmetry, the HOMO-LUMO transition is allowed. For oxaadamantane the slowly but steadily increasing absorption signal is responsible for the comparably low absorption onset. Another explanation could be a deviation of oxaadamantane rather than oxadiamantane. The absorption behavior of oxaadamantane differs at low energies from those of the other two oxadiamondoids. This is also apparent from the integrated absorption signal in the right panel of Fig. 6.20.

The absorption features and the optical gap exhibit a clear size dependence in oxadiamondoids. This is unlike the observation for diamondoid thiols which exhibit no size dependence of the absorption inset. The size dependent effects, however, are slightly smaller than in diamondoids. Quantum chemical calculations have been conducted to derive a qualitative understanding of the effect of the oxygen-incorporation on the electronic structure of the diamondoids.[8] The same methods that were successfully used to describe the optical properties of pristine diamondoids [96], however, failed to reproduce the optical spectra of oxadiamondoids [133]. Therefore, the following theoretical investigations are aimed only at a qualitative understanding.

The highest occupied and lowest unoccupied molecular orbitals of oxaadamantane and oxadiamantane are shown in Fig. 6.21 together with the comparable orbitals of adamantane. First of all it can be noticed that the molecular orbitals of adamantane and oxaadamantane, in general, are very similar in shape and electron distribution. For the unoccupied orbitals large fraction of the electron density are delocalized on the hydrogen surface while the highest occupied orbitals are localized inside the cluster exhibiting notable σ-character. The oxygen atom in both oxaadamantane and oxadiamantane is not involved in the formation

---

[8]The oxadiamondoids have been geometry optimized at 6-311++G** level of theory using the B3LYP hybrid functional as implemented in TURBOMOLE [71] and their molecular orbitals were computed.

6.3. BULK-SUBSTITUTED DIAMONDOIDS

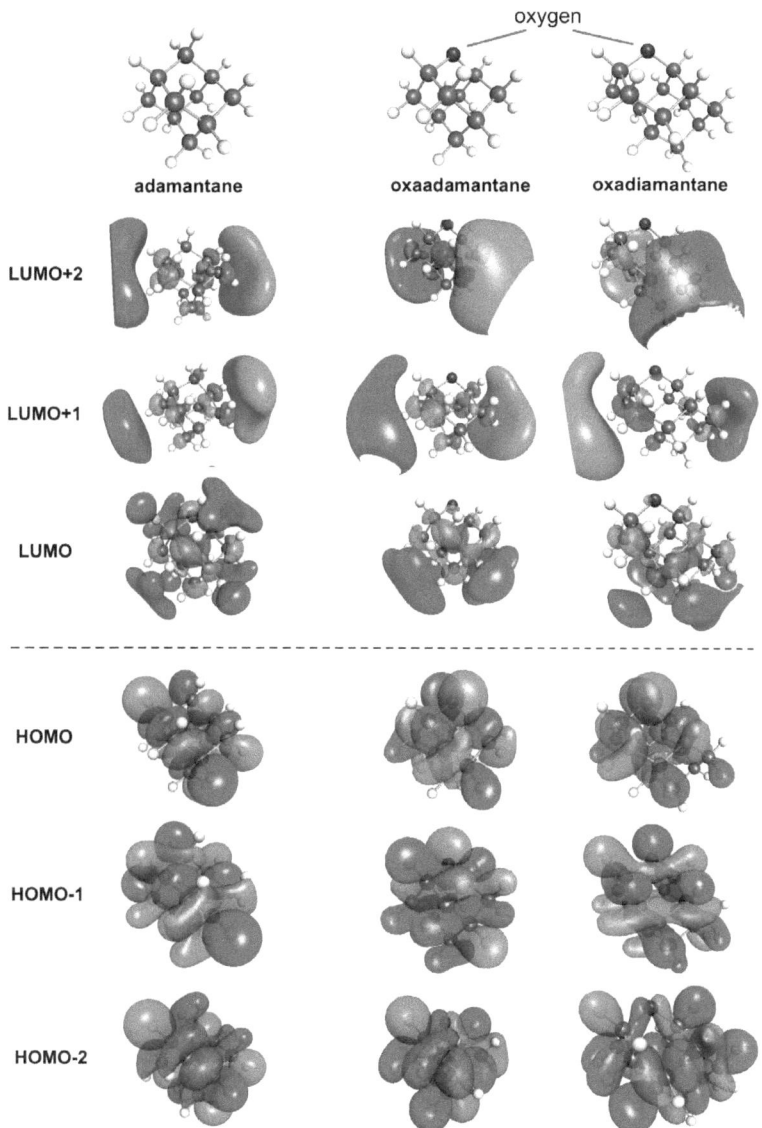

Figure 6.21: The molecular orbitals of adamantane, oxaadamantane, and oxadiamantane. The occupied orbitals of the oxadiamondoids are similar to those of the pristine species but exhibit a variable contribution of oxygen atom. The influence of the oxygen is strongest in the HOMO. The incorporated oxygen does not participate in any of the unoccupied orbitals. Similar behavior can be observed for the orbitals of oxatriamantane which are not shown.

of the lowest unoccupied molecular orbitals. The LUMO+1 and LUMO+2 are very similar to those of the pristine diamondoid. In the LUMO the similarity is limited to the side of the oxadiamondoid which lies opposite the oxygen.

For the occupied orbitals the situation is different. Fig. 6.21 shows that the HOMO of the oxadiamondoids exhibits a large contribution from the oxygen atom. Comparison with the HOMO of adamantane demonstrates that the oxygen contribution mixes strongly with the diamondoid's HOMO wavefunction. The same is true for the HOMO−1 where the oxygen contribution is less pronounced and similar in magnitude to the regular diamondoid's wavefunction. In the HOMO−2 the participation of the incorporated oxygen in the formation of the orbital wanes.

The recorded spectra in combination with the computed orbitals suggest that the incorporation of an oxygen atom into the diamondoid framework leads to a mixing of the diamondoid wavefunctions with the electronic states which are localized near the oxygen. The impurity atom is interwoven in the electronic structure of the diamondoid and only slightly alters the nature of the electronic levels of the diamondoid. Additional molecular orbitals close to the band edge or even electronic states that have the character of impurities are not observed. These observations are in stark contrast to those for diamondoid thiols where the functional group leads to an impurity state in the band gap but otherwise leaves the electronic structure of the diamondoid nearly unchanged. However, similar to the case of diamondoid thiols, no detectable photoluminescence was found for oxadiamondoids. An investigation of the band edges of oxadiamondoids, in particular of the occupied states, would help to refine the understanding of the changes in the electronic structure.

Nitrogen

Another interesting substituent that is of particular importance for diamond materials is nitrogen. It is a common impurity and a dopant in diamond materials. It is well known through its occurrence in the form of nitrogen-vacancy (NV) centers which are used to introduce visible luminescence and, most notably, single photon emission [134]. NV centers in diamond are promising candidates for various applications ranging from single photon sources for quantum cryptography to biolabelling. Reducing the size of diamonds which contain these NV centers would be highly favorable for many of the considered applications. E.g.,

6.3. BULK-SUBSTITUTED DIAMONDOIDS

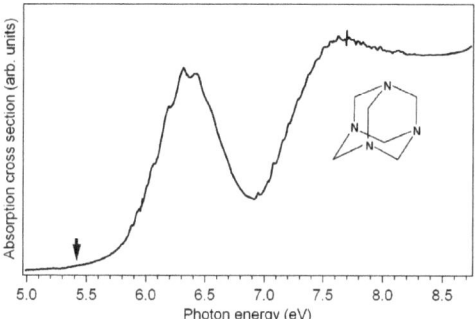

Figure 6.22: Optical absorption of urotropine ($C_6N_4H_{12}$, also hexamethylenetetramine) measured in the gas phase at comparable absorption (T=390 K). The optical gap at 5.42 eV, as defined in section 5.1.3, is marked by an arrow.

in quantum magnetometry [135, 136] single spins are detected with a sensitivity that is inversely proportional to the cube of the distance between the sensor (i.e., the NV center) and the spin which is detected. Diamondoids with sizes below 1 nm, however, are far too small to incorporate vacancies in their lattice structure. The smallest nanodiamonds to contain NV centers reported to date still measure 5 nm in diameter [40]. However, with the precise structural control offered by diamondoids this size is likely not to be the absolute lower limit.

As a first step towards a better understanding of nitrogen impurities in a diamond lattice urotropine is investigated. Urotropine ($C_6N_4H_{12}$, also called hexamethylenetetramine) consists of an adamantane cage in which the four tertiary carbon atoms are replaced by nitrogen. All remaining carbon atoms are still passivated with hydrogen while the nitrogen is only bonded to carbon.

The optical absorption spectrum of urotropine is shown in Fig. 6.22. The spectrum is found to be rather smooth and sharp spectral features, as observed for adamantane in section 5.1, are absent. The absorption onset is very smooth and the optical gap, as defined in section 5.1.3, is determined to be 5.42±0.04 eV, about 1 eV lower than for adamantane. Similar to adamantane, the spectrum exhibits two broad main bands. The first one ranges from the absorption onset to approx. 7 eV and the second starts at 7 eV and fades into the ionization continuum beyond 8 eV. The urotropine structure corresponds to the inclusion of four nitrogen atoms into the adamantane framework. The additional valence electrons of the N atoms lead to electron lone pairs replacing covalently bonded hydrogen. These electron lone pairs constitute the highest occupied states [33] which are broad in energy and dominate the optical properties near the gap resulting in a lower optical gap and a smooth appearance of the spectrum. As for the oxygen inclusions, no UV photoluminescence from urotropine is observed upon excitation between 5.6 and 8.8 eV. This lack of UV luminescence suggests that the incorporation of impurity atoms favors nonradiative decay.

Chapter 7

Summary & Outlook

In this work the optical properties of pristine and chemically modified diamondoids have been studied. The quality of the experimental data allows for a direct comparison to computational approaches to the electronic structure of diamondoids. These data serve as a benchmark for theoretical investigation and have already triggered first follow-up studies that deepened the understanding of the underlying effects.

The absorption data of eleven different diamondoid structures have been recorded. The data allowed for the first time to separate the influence of size and shape on a nanocrystal's optical properties. The presented data reveal a strong influence of the diamondoid's *shape* on its optical response which even outweighs size effects. The different diamondoid structures were categorized on the basis of their optical response and their geometry in 1D, 2D and 3D structures. Further it was shown that the characteristic optical response of bulk diamond evolves within the 3D structural family already on a sub-nanometer scale. This showed that the smallest diamond to exhibit bulk-like optical properties is defined by shape rather than size.

Within this work photoluminescence from diamondoids was discovered for the first time and energy resolved spectra were recorded for several diamondoid species. The emitted light is broad in energy and lies in the ultraviolet spectral region where alternative photonic materials are scarce. The quantum yield, however, was found to be low. The analysis of the structure in the photoluminescence spectra lead to an explanation based on a severe distortion of the nanocrystal in the excited state.

The photoluminescence properties and even more the absorption behavior were found to strongly depend on the shape of the diamondoid. The case of the optical absorption, where the effect of shape even outweighs size effects, demonstrates that shape becomes a non-negligible design parameter in nanocrystal on a (sub-)nanometer length scale. This implies the possibility, if not the necessity, of engineering the optical properties of (sub-)nanocrystals by controlling their shape. Shape as a design parameter could give rise to a new field of *shape-dependent photophysics*.

As a second endeavor the influence of different kinds of impurities on the optical properties of diamondoids were investigated. The influence of surface functional groups, using the technologically relevant example of thiol groups, was studied as well as the effect of incorporating impurity atoms in the diamondoid's carbon framework.

For functionalization it was shown that a thiol group that is attached to the diamond cluster completely dominates the optical properties. Upon thiolation the spectra of all diamondoids become similarly smooth and the optical gap loses its size dependence. The origin of the drastic changes was identified as what could be called an impurity state which lies just above the valence states of the diamondoids. This state, which is fix in energy even for different diamondoid size due to its localized nature, could clearly be distinguished from the diamondoid's electronic structure. The thiol group is also found to quench the photoluminescence in diamondoids. In combination with theoretical investigations the origin of the quenching was identified to lie in a dissociative excitation of the S-H bond. Further theoretical and photoelectron studies were conducted on larger diamondoids which suggest that photoluminescence is likely to return in larger diamondoid thiols.

The incorporation of an impurity atom was shown to have a notably different effect on the electronic structure of diamondoids. For the example of oxadiamondoids the size dependence of the absorption is slightly diminished but persists in principle. Rather than adding new isolated energy levels to the electronic structure of the diamondoid, the molecular orbitals of oxadiamondoids are found to be combinations of the diamondoid orbitals with variable contributions of the oxygen levels.

The investigation of modified diamondoids showed that two different approaches exist that lead to fundamentally distinct changes in the electronic structure: The bulk-substitution with oxygen leads to an incorporation of the substituent into the electronic structure of the diamondoid while the thiol functional group adds an isolated electronic level - an *impurity level* - to the system.

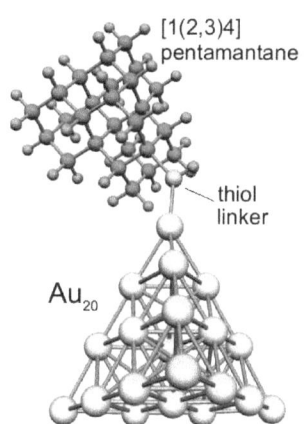

Figure 7.1: Optimized structure of a diamondoid-Au$_{20}$ cluster hybrid. [137]

Diamondoids display unique electronic and optical properties. Further advances in diamondoid chemistry will broaden the range of possible applications in the future. For example, the optical gap of diamondoids could be lowered to the near UV spectral region or even into the visible regime by *push-pull* doping, i.e., through targeted modification which increases the HOMO and lowers the LUMO energies. Such an approach could use two different functional groups or a suitable combination of functional groups and bulk-substituents to lower the gap energy.

The combination of diamondoids with other nanocarbon materials such as carbon nanotubes or graphene would allow one to combine their unique properties. These materials could rely on the strength of the covalent carbon-carbon bond which would ensure a high structural stability. Further, the use of functional group linkers could be circumvented and thus this integrated *all-carbon* approach would not have to struggle with compatibility issues. A pure carbon "hybrid" of diamondoids and carbon nanotubes (or graphene) could for example serve as a nano-field emitting device in which electrons that are delivered by the conducting nanotube are emitted via the negative electron affinity surface of the diamondoid. Nanophotonic devices could be realized in a similar fashion. Such mixed sp^2/sp^3 nanocarbon devices could initiate the interesting field of *carbon nanoelectronics*.

Another interesting possibility are hybrid cluster systems which could be made up of, e.g., a diamondoid thiol and a gold cluster shown in Fig. 7.1. Experimental investigations of these cluster hybrids could give rise to new interesting materials and also foster our understanding of hybrid systems in general. They would, e.g., allow a detailed experimental investigation of the nature of the S-Au bond.

The progress in the field will critically depend on the availability of diamondoid materials. The large scale synthesis of higher diamondoids, e.g., by means of standard CVD diamond growth methods, will be necessary to provide diamondoid materials to the scientific community and to further foster the growing technological interest in diamondoid electronics.

Appendix

Appendix A

Character Tables and Selection Rules

Appendix A. Character Tables and Selection Rules

In the following the character tables of all relevant point groups are listed starting with the lowest symmetry. Below each character table the dipole-forbidden transitions are listed. They are derived as explained in detail in section 3.3. Also the transformation behavior of x, y, and z are given. As a convention the z-direction points along the main rotational axis.

The C_s Point Group:
Apical Thiols incl. Adamantane-1-thiol & Oxadiamantane

C_2	E	σ_h
A'	1	1
A''	1	-1

Table A.1: Character table of the C_s point group. Within this group x and y transform as A' and z as A''.

	$A' \triangleq x, y$	$A'' \triangleq z$	Forbidden transitions
A':	$A' \otimes A' = A'$	$A' \otimes A'' = A''$	none
A'':	$A'' \otimes A' = A''$	$A'' \otimes A'' = A'$	

The C_2 Point Group: [123] **Tetramantane**

C_2	E	C_2
A	1	1
B	1	-1

Table A.2: Character table of the C_2 point group. Within this group x and y transform as B and z as A.

	$A \triangleq z$	$B \triangleq x, y$	Forbidden transitions
A:	$A \otimes A = A$	$A \otimes B = B$	none
B:	$B \otimes A = B$	$B \otimes B = A$	

APPENDIX A. CHARACTER TABLES AND SELECTION RULES

The C_{2h} Point Group: [121] Tetramantane

C_{2h}	E	C_2	i	σ_h
A_g	1	1	1	1
A_u	1	1	-1	-1
B_g	1	-1	1	-1
B_u	1	-1	-1	1

Table A.3: Character table of the C_{2h} point group. Within this group x and y transform as B_u and z transforms as A_u.

	$A_u \stackrel{\wedge}{=} z$	$B_u \stackrel{\wedge}{=} x, y$	Forbidden transitions
A_g:	$A_g \otimes A_u = A_u$	$A_g \otimes B_u = B_u$	$A_g \to g$
A_u:	$A_u \otimes A_u = A_g$	$A_u \otimes B_u = B_g$	$A_u \to u$
B_g:	$B_g \otimes A_u = B_u$	$B_g \otimes B_u = A_u$	$B_g \to g$
B_u:	$B_u \otimes A_u = B_g$	$B_u \otimes B_u = A_g$	$B_u \to u$

The letters g and u stand short for the set of all *even* and *odd* representations, respectively.

The C_{2v} Point Group:
Triamantane & [1212] Pentamantane, Oxaadamantane & Oxatriamantane

C_{2v}	E	C_2	σ_v	σ_v'
A_1	1	1	1	1
A_2	1	1	-1	-1
B_1	1	-1	1	-1
B_2	1	-1	-1	1

Table A.4: Character table of the C_{2v} point group. Within this group x transforms as B_1, y as B_2, and z as A_1.

	$A_1 \stackrel{\wedge}{=} z$	$B_1 \stackrel{\wedge}{=} x$	$B_2 \stackrel{\wedge}{=} y$	Forbidden transitions
A_1:	$A_1 \otimes A_1 = A_1$	$A_1 \otimes B_1 = B_1$	$A_1 \otimes B_2 = B_2$	$A_1 \to A_2$
A_2:	$A_2 \otimes A_1 = A_2$	$A_2 \otimes B_1 = B_2$	$A_2 \otimes B_2 = B_1$	$A_2 \to A_1$
B_1:	$B_1 \otimes A_1 = B_1$	$B_1 \otimes B_1 = A_1$	$B_1 \otimes B_2 = A_2$	$B_1 \to B_2$
B_2:	$B_2 \otimes A_1 = B_2$	$B_2 \otimes B_1 = A_2$	$B_2 \otimes B_2 = A_1$	$B_2 \to B_1$

The C_{3v} Point Group: [1(2)3] Tetramantane

C_{3v}	E	$2C_3$	$3\sigma_v$
A_1	1	1	1
A_2	1	1	-1
E	2	-1	0

Table A.5: Character table of the C_{3v} point group. Within this group x and y transform as E and z as A_1.

	$A_1 \stackrel{\wedge}{=} z$	$E \stackrel{\wedge}{=} x, y$	Forbidden transitions
A_1:	$A_1 \otimes A_1 = A_1$	$A_1 \otimes E = E$	$A_1 \to A_2$
A_2:	$A_2 \otimes A_1 = A_2$	$A_2 \otimes E = E$	$A_2 \to A_1$
E:	$E \otimes A_1 = E$	$E \otimes E = A_1 \oplus A_2 \oplus E$	–

The D_{3d} Point Group: Diamantane & [12312] Hexamantane

D_{3d}	E	$2C_3$	$3C_2'$	i	$2S_6$	$3\sigma_d$
A_{1g}	1	1	1	1	1	1
A_{2g}	1	1	-1	1	1	-1
A_{1u}	1	1	1	-1	-1	-1
A_{2u}	1	1	-1	-1	-1	1
E_g	2	-1	0	2	-1	0
E_u	2	-1	0	-2	1	0

Table A.6: Character table of the D_{3d} point group. Within this group x and y transform as E_u and z as A_{2u}.

Appendix A. Character Tables and Selection Rules

	$A_{2u} \stackrel{\wedge}{=} z$	$E_u \stackrel{\wedge}{=} x,y$	Forbidden transitions
A_{1g}:	$A_{1g} \otimes A_{2u} = A_{2u}$	$A_{1g} \otimes E_u = E_u$	$A_{1g} \to g, A_{1u}$
A_{2g}:	$A_{2g} \otimes A_{2u} = A_{1u}$	$A_{2g} \otimes E_u = E_u$	$A_{2g} \to g, A_{2u}$
A_{1u}:	$A_{1u} \otimes A_{2u} = A_{2g}$	$A_{1u} \otimes E_u = E_g$	$A_{1u} \to u, A_{1g}$
A_{2u}:	$A_{2u} \otimes A_{2u} = A_{1g}$	$A_{2u} \otimes E_u = E_g$	$A_{2u} \to u, A_{2g}$
E_g:	$E_g \otimes A_{2u} = E_u$	$E_g \otimes E_u = A_{1u} \oplus A_{2u} \oplus E_u$	$E_g \to g$
E_u:	$E_u \otimes A_{2u} = E_g$	$E_u \otimes E_u = A_{1g} \oplus A_{2g} \oplus E_g$	$E_u \to u$

The letters g and u stand short for the set of all *even* and *odd* representations, respectively.

The T_d Point Group: Adamantane & $[1(2,3)4]$ Pentamantane

T_d	E	$8C_3$	$3C_2'$	$6S_4$	$6\sigma_d$
A_1	1	1	1	1	1
A_2	1	1	1	-1	-1
E	2	-1	2	0	0
T_1	3	0	-1	1	-1
T_2	3	0	-1	-1	1

Table A.7: Character table of the T_d point group. Within this group x, y, and z transform as T_2.

	$T_2 \stackrel{\wedge}{=} x,y,z$	Forbidden transitions
A_1:	$A_1 \otimes T_2 = T_2$	$A_1 \to A_1, A_2, E, T_1$
A_2:	$A_2 \otimes T_2 = T_1$	$A_2 \to A_1, A_2, E, T_2$
E:	$E \otimes T_2 = T_1 \oplus T_2$	$E \to A_1, A_2, E$
T_1:	$T_1 \otimes T_2 = A_2 \oplus E \oplus T_1 \oplus T_2$	$T_1 \to A_1$
T_2:	$T_2 \otimes T_2 = A_1 \oplus E \oplus T_1 \oplus T_2$	$T_2 \to A_2$

Appendix B

List of Publications

Publications

Excerpts of the present work have formerly been published in the following form:

Optical Response of Diamond Nanocrystals as a Function of Particle Size, Shape, and Symmetry
L. Landt, K. Klünder, J. E. Dahl, R. M. K. Carlson, T. Möller, and C. Bostedt, Phys. Rev. Lett. **103**, 047402 (2009).

Intrinsic photoluminescence of adamantane in the ultraviolet spectral region
L. Landt, W. Kielich, D. Wolter, M. Staiger, A. Ehresmann, T. Möller, and C. Bostedt, Phys. Rev. B **80**, 205323 (2009).

The influence of a single thiol group on the electronic and optical properties of the smallest diamondoid adamantane
L. Landt, M. Staiger, D. Wolter, K. Klünder, P. Zimmermann, T. M. Willey, T. van Buuren, D. Brehmer, P. R. Schreiner, B. A. Tkachenko, A. A. Fokin, T. Möller, and C. Bostedt, J. Chem. Phys. **132**, 024710 (2010).

Experimental and theoretical study of the absorption properties of thiolated diamondoids
L. Landt, C. Bostedt, D. Wolter, T. Möller, J. E. Dahl, R. M. K. Carlson, B. A. Tkachenko, A. A. Fokin, P. R. Schreiner, A. Kulesza, R. Mitrić, and V. Bonačić-Koutecký, J. Chem. Phys. **132**, 144305 (2010).

Diamondoids
C. Bostedt, **L. Landt**, T. Möller, J. E. Dahl, R. M. K. Carlson,
to appear in *Nature's Nanostructures*, H. Guo and A. S. Barnard (Eds.), Pan Stanford Publishing

Related own work:

Experimental determination of the ionization potentials of the first five members of the nanodiamond series
K. Lenzke, **L. Landt**, M. Hoener, H. Thomas, J. E. Dahl, S. G. Liu, R. M. K. Carlson, T. Möller, and C. Bostedt, J. Chem. Phys. **127**, 084320 (2007).

Separation of size-dependent screening effects and quantum confinement of the occupied states of diamondoids
L. Landt, K. Klünder, T. M. Willey, T. van Buuren, J. E. Dahl, R. M. K. Carlson, T. Möller, and C. Bostedt, *in preparation.*

List of Tables

2.1	Structural properties of diamondoids	8
2.2	Ionization potentials of diamondoids	14
3.1	Molecular point groups	35
4.1	PL - Measuring parameters	59
4.2	UPS - Experimental parameters	65
5.1	Optical gaps of diamondoids	85
5.2	Dipole selection rules for diamondoids	91
5.3	Photoluminescence characteristics of diamondoids	115
6.1	Electronic key figures of adamantane-1-thiol	135

Bibliography

[1] J. E. Dahl, S. G. Liu, and R. M. K. Carlson, Science **299**, 96 (2003).

[2] W. L. Yang, J. D. Fabbri, T. M. Willey, J. R. I. Lee, J. E. Dahl, R. M. K. Carlson, P. R. Schreiner, A. A. Fokin, B. A. Tkachenko, N. A. Fokina, W. Meevasana, N. Mannella, K. Tanaka, X. J. Zhou, T. van Buuren, M. A. Kelly, Z. Hussain, N. A. Melosh, and Z.-X. Shen, Science **316**, 1460 (2007).

[3] Y. Lifshitz, T. Köhler, T. Frauenheimer, I. Guzmann, A. Hoffmann, R. Q. Zhang, X. T. Zhou, and S. T. Lee, Science **297**, 1531 (2002).

[4] H. Schwertfeger, A. A. Fokin, and P. R. Schreiner, Angew. Chem. Int. Ed. **47**, 1022-1036 (2008).

[5] M. Staiger, Photoelektronenspektroskopie an funktionalisierten Nanodiamanten, Diplomarbeit, Berlin 2009.

[6] S. Landa and V. Machnacek, Collect. Czech. Chem. Commun. **5** (1933).

[7] V. Prelog and R. Seiwerth, Ber. **74**, 1644 (1941).

[8] P. von R. Schleyer, J. Am. Chem. Soc. **79**, 3202 (1957).

[9] C. Cupas, P. von R. Schleyer, and D. J. Trecker, J. Am. Chem. Soc. **87**, 917 (1965).

[10] O. Vogl and B. C. Anderson, Tetrahedron Letters **4**, 415 (1966).

[11] J. W. van Zandt, P. von R. Schleyer, G. J. Gleicher, and L. B. Rodewald, J. Am. Chem. Soc. **88**, 3862 (1966).

[12] S. Hála and S. Landa, Angew. Chem. internat. Edit. **5**, 1045 (1966).

[13] M. A. McKervey, Tetrahedron **36**, 971 (1980).

[14] W. S. Wingert, Fuel **71**, 37 (1992).

[15] R. Lin and Z. A. Wilk, Fuel **74**, 1512 (1995).

[16] A. Shimoyama and H. Yabuta, Geochem. J. **36**, 173 (2002).

[17] J. Chen, J. Fu, G. Sheng, D. Liu, and J. Zhang, Org. Geochem. **25**, 179 (1996).

[18] K. Grice, R. Alexander, and R. I. Kagi, Org. Geochem. **31**, 67 (2000).

[19] J. E. Dahl, J. M. Moldowan, K. E. Peters, G. E. Claypool, M. A. Rooney, G. E. Micheal, M. R. Mello, and M. L. Kohnen, Nature **399**, 54 (1999).

[20] K. Lenzke, Photoionisation von molekularen Diamanten, Diplomarbeit, Berlin 2006.

[21] Chevron Ventures, MolecularDiamond Technologies, product sheet.

[22] J. Reiser, E. McGregor, J. Jones, R. Enick, and G. Holder, Fluid Phase Equilibria **117**, 160 (1996).

[23] A. T. Balaban and P. von R. Schleyer, Tetrahedron **34**, 3599 (1978).

[24] T. M. Willey, C. Bostedt, T. van Buuren, J. E. Dahl, S. G. Liu, R. M. K. Carlson, L. J. Terminello, and T. Möller, Phys. Rev. Lett. **95**, 113401 (2005).

[25] K. Lenzke, L. Landt, M. Hoener, H. Thomas, J. E. Dahl, S. G. Liu, R. M. K. Carlson, T. Möller, and C. Bostedt, J. Chem. Phys. **127**, 084320 (2007).

[26] K. Klünder, Photoelektronenspektroskopie an molekularen Diamanten, Diplomarbeit, Berlin 2007.

[27] J. Filik, J. N. Harvey, N. L. Allan, P. W. May, J. E. P. Dahl, S. Liu, and R. M. K. Carlson, Spectrochim. Acta Part A **64**, 681 (2005).

[28] J. Oomens, N. Polfer, O. Pirali, Y. Ueno, R. Maboudian, P. W. May, J. Filik, J. E. Dahl, S. Liu, and R. M. K. Carlson, J. Mol. Spec. **238**, 158 (2006).

[29] G. C. McIntosh, M. Yoon, S. Berber, and D. Tománek, Phys. Rev. B **70**, 045401 (2004).

[30] A. J. Lu, B. C. Pan, and J. G. Han, Phys. Rev. B **72**, 035447 (2005).

[31] N. D. Drummond, A. J. Williamson, R. J. Needs, and G. Galli, Phys. Rev. Lett. **95**, 096801 (2005).

[32] J. W. Raymonda, J. Chem. Phys. **56**, 3912 (1972).

[33] W. Schmidt, Tetrahedron **29**, 2129 (1973).

[34] T. van Buuren, L. N. Dinh, L. L. Chase, W. J. Siekhaus, and L. J. Terminello, Phys. Rev. Lett. **80**, 3803 (1998).

[35] C. Bostedt, T. van Buuren, T. M. Willey, N. Franco, L. J. Terminello, C. Heske, and T. Möller, Appl. Phys. Lett. **84**, 4056 (2004).

[36] T. M. Willey, C. Bostedt, T. van Buuren, J. E. Dahl, S. G. Liu, R. M. K. Carlson, R. W. Meulenberg, E. J. Nelson, and L. J. Terminello, Phys. Rev. B **74**, 205432 (2006).

[37] L. Landt, K. Klünder, T. M. Willey, T. van Buuren, J. E. Dahl, R. M. K. Carlson, T. Möller, and C. Bostedt, in preparation .

[38] The Nobel Prize committe of The Royal Swedish Academy of Sciences, http://nobelprize.org/.

[39] T. D. Ladd, F. Jelezko, R. Laflamme, Y. Nakamura, C. Monroe, and J. L. O'Brien, Nature (2010).

[40] B. R. Smith, D. W. Inglis, B. Sandnes, J. R. Rabeau, A. V. Zvyagin, D. Gruber, C. J. Noble, R. Vogel, E. Osawa, and T. Plakhotnik, small **5**, 1649 (2009).

[41] P. R. Schreiner, N. A. Fokina, B. A. Tkachenko, H. Hausmann, M. Serafin, J. E. P. Dahl, S. Liu, R. M. K. Carlson, and A. A. Fokin, J. Org. Chem. **71**, 6709 (2006).

[42] Y. Wang, E. Kioupakis, X. Lu, D. Wegner, R. Yamachika, J. E. Dahl, R. M. K. Carlson, S. G. Louie, and M. F. Crommie, Nature Materials **7**, 38 (2008).

[43] A. A. Fokin and P. R. Schreiner, Mol. Phys. **107**, 823 (2009).

[44] A. P. Marchand, Science **299**, 52 (2003).

[45] N. D. Drummond, Nat. Nanotech. **2**, 462 (2007).

[46] W. A. Clay, Z. Liu, W. L. Yang, J. D. Fabbri, J. E. Dahl, R. M. K. Carlson, Y. Sun, P. R. Schreiner, A. A. Fokin, B. A. Tkachenko, N. A. Fokina, P. A. Pianetta, N. Melosh, and Z.-X. Shen, Nano Lett. **9**, 57 (2009).

[47] W. J. Parak, D. Gerion, D. Zanchet, A. S. Woerz, T. Pellegrino, C. Micheel, S. C. Williams, M. Seitz, R. E. Bruehl, Z. Bryant, C. Bustamante, C. R. Bertozzi, and A. P. Alivisatos, Chem. Mater. **14**, 2113 (2002).

[48] M. F. Calhoun, J. Sanchez, D. Olaya, M. E. Gershenson, and V. Podzorov, Nature Materials **7**, 84 (2008).

[49] J. C. Love, L. A. Estroff, J. K. Kriebel, R. G. Nuzzo, and G. M. Whitesides, Chem. Rev. **105**, 1103 (2005).

[50] S. K. Arya, P. R. Solanki, M. Datta, and B. D. Malhotra, Biosensors and Bioelectronics **24**, 2810 (2009).

[51] Lawrence Berkeley Laboratory, Advanced Light Source, http://www.als.lbl.gov.

[52] A. A. Fokin, P. R. Schreiner, N. A. Fokina, B. A. Tkachenko, H. Hausmann, M. Serafin, J. E. P. Dahl, S. Liu, and R. M. K. Carlson, J. Org. Chem. **71**, 8532 (2006).

[53] A. A. Fokin, E. D. Butova, L. V. Chernish, N. A. Fokina, J. E. P. Dahl, R. M. K. Carlson, and P. R. Schreiner, Org. Lett. **9**, 2541 (2007).

[54] H. Schwertfeger, C. Würtele, H. Hausmann, J. E. P. Dahl, R. M. K. Carlson, A. A. Fokin, and P. R. Schreiner, Adv. Synth. Catal. **351**, 1041 (2009).

[55] A. A. Fokin, T. S. Zhuk, A. E. Pashenko, P. O. Dral, P. A. Gunchenko, J. E. P. Dahl, R. M. K. Carlson, T. V. Koso, M. Serafin, and P. R. Schreiner, Org. Lett. **11**, 3068 (2009).

[56] F. Marsusi and K. Mirabbaszadeh, J. Phys.: Condens. Matter **21**, 215303 (2009).

[57] J. C. Garcia, J. F. Justo, W. V. M. Machado, and L. V. C. Assali, Phys. Rev. B **80**, 125421 (2009).

[58] B. A. Tkachenko, N. A. Fokina, L. V. Chernish, J. E. P. Dahl, S. Liu, R. M. K. Carlson, A. A. Fokin, and P. R. Schreiner, Org. Lett. **8**, 1767 (2005).

[59] T. M. Willey, J. D. Fabbri, J. R. I. Lee, P. R. Schreiner, A. A. Fokin, B. A. Tkachenko, N. A. Fokina, J. E. Dahl, R. M. K. Carlson, A. L. Vance, W. Yang, L. J. Terminello, T. van Buuren, and N. A. Melosh, J. Am. Chem. Soc. **130**, 10536 (2008).

[60] W. Zhang, B. Gao, J. Yang, V. Carvetta, and Y. Luo, J. Chem. Phys. **130**, 054705 (2009).

[61] S. Roth, D. Leuenberger, J. Osterwalder, J. E. Dahl, R. M. K. Carlson, B. A. Tkachenko, A. A. Fokin, P. R. Schreiner, and M. Hengsberger, Chem. Phys. Lett. **495**, 102 (2010).

[62] J. Murrell, S. Kettle, and J. Tedder, *The Chemical Bond*, Wiley, 1985.

[63] R. Woolfson and J. M. Paschoff, *Physics*, Scott, Foresman and Company, 1990.

[64] P. Y. Yu and M. Cardona, *Fundamentals of Semiconductors*, Graduate Texts in Physics, Springer, 2010.

[65] C. Kittel, *Introduction to Solid State Physics*, John Wiley & Sons, Inc., 1986.

[66] C. Bostedt, Electronic structure of germanium nanocrystal films probed with synchrotron radiation, Dissertation, 2002.

[67] M. S. Hybertsen, Phys. Rev. Lett. **72**, 1514 (1994).

[68] D. Kovalev, H. Heckler, M. Ben-Chorin, G. Polisski, M. Schwartzkopff, and F. Koch, Phys. Rev. Lett. **81**, 2803 (1998).

[69] L. T. Canham, Appl. Phys. Lett. **57**, 1046 (1990).

[70] M. J. Frisch et al., Gaussian 03, Revision C.02, Gaussian, Inc., Wallingford, CT, 2004.

[71] TURBOMOLE V6.1 2009, a development of University of Karlsruhe and Forschungszentrum Karlsruhe GmbH, 1989-2007, TURBOMOLE GmbH, since 2007; available from http://www.turbomole.com.

[72] W. J. Hehre, *Ab initio molecular orbital theory*, Wiley, 1986.

[73] C. Lee, W. Yang, and R. G. Parr, Phys. Rev. B **37**, 785 (1988).

[74] J. W. Leech and D. J. Newman, *How To Use Groups*, Methuen's Monographs on Physical Subjects, Science Paperbacks, 1969.

[75] M. Wagner, *Gruppentheoretische Methoden in der Physik*, Vieweg, 1998.

[76] C. Bostedt, T. van Buuren, T. M. Willey, and L. J. Terminello, Appl. Phys. Lett. **85**, 5334 (2004).

[77] Sigma-Aldrich, http://www.sigmaaldrich.com/.

[78] P. R. Schreiner, private communication.

[79] L. Landt, Elektronische Struktur und Absorption von Nanodiamanten, Diplomarbeit, Berlin 2006.

[80] Korth Kristalle GmbH, http://www.korth.de.

[81] A. Schlachter and F. Wuilleumiers, *New Direction in Research with Third-Generation Soft X-Ray Synchrotron Sources*, Kluwer Academic Publishers, 1994.

[82] J. Falta and T. Möller, *Forschung mit Synchrotronstrahlung*, Vieweg + Teubner, 2010.

[83] M. Joppien, Lumineszenzspektroskopische Untersuchungen der elektronischen Anregungen von Helium- und Neon-Clustern, Dissertation, Hamburg 1994.

[84] D. Wolter, Dissertation, in progress.

[85] A. Ehresmann, L. Werner, S-Klumpp, P. V. Demekhin, M. P. Lemeshko, V. L. Sukhorukov, K.-H. Schartner, and H. Schmoranzer, J. Phys. B: At. Mol. Opt. Phys. **39**, L119 (2006).

[86] W. Kielich, private communication.

[87] A. Potts and W. Price, Proc. Roy. Soc. Lond. A **326**, 181 (1972).

[88] National Institute of Standards and Technology, http://webbook.nist.gov/.

[89] P. Denham, E. C. Lightowlers, and P. J. Dean, Phys. Rev. **161**, 762 (1967).

[90] L. Landt, K. Klünder, J. E. Dahl, R. M. K. Carlson, T. Möller, and C. Bostedt, Phys. Rev. Lett. **103**, 047402 (2009).

[91] P. J. Dean and J. C. Male, Proc. Roy. Soc. A **277**, 330 (1964).

[92] R. Schäfer and J. A. Becker, Phys. Rev. B **54**, 10296 (1996).

[93] I. Vasiliev, S. Öğüt, and J. R. Chelikowsky, Phys. Rev. Lett. **86**, 1813 (2001).

[94] J.-Y. Raty, G. Galli, C. Bostedt, T. W. van Buuren, and L. J. Terminello, Phys. Rev. Lett. **90**, 037401 (2003).

[95] G. D. Scholes and G. Rumbles, Nat. Mat. **5**, 683 (2006).

[96] M. Vörös and A. Gali, Phys. Rev. B **80**, 161411(R) (2009).

[97] J.-Y. Raty and G. Galli, J. Electroanal. Chem. **584**, 9 (2005).

[98] M. Rohlfing and S. G. Louie, Phys. Rev. Lett. **80**, 3320 (1998).

[99] L. X. Benedict, A. Puzder, A. J. Williamson, J. C. Grossman, G. Galli, J. E. Klepeis, J.-Y. Raty, and O. Pankratov, Phys. Rev. B **68**, 085310 (2003).

[100] F. Priolo, G. Franyò, D. Pacifici, V. Vinciguerra, F. Iacona, and A. Irrera, J. Appl. Phys. **89**, 264 (2001).

[101] L. Landt, W. Kielich, D. Wolter, M. Staiger, A. Ehresmann, T. Möller, and C. Bostedt, Phys. Rev. B **80**, 205323 (2009).

[102] F. Hirayama and S. Lipsky, J. Chem. Phys. **51**, 3616 (1969).

[103] F. Hirayama and S. Lipsky, Chem. Phys. Lett. **22**, 172 (1973).

[104] F. Hirayama, W. Rothman, and S. Lipsky, Chem. Phys. Lett. **5**, 296 (1970).

[105] W. Rothman, F. Hirayama, and S. Lipsky, J. Chem Phys. **58**, 1300 (1973).

[106] Y. Kanemitsu and S. Okamoto, Phys. Rev. B **58**, 9652 (1998).

[107] M. Nirmal and L. Brus, Acc. Chem. Res. **32**, 407 (1999).

[108] G. Allan, C. Delerue, and M. Lannoo, Phys. Rev. Lett. **76**, 2961 (1996).

[109] V. A. Belyakov, V. A. Burdov, R. Lockwood, and A. Meldrum, Adv. Opt. Tech. **2008**, 279502 (2008).

[110] T. J. Mullen, P. Zhang, C. Srinivasan, M. W. Horn, and P. S. Weiss, J. Electroanal. Chem. **621**, 229 (2008).

[111] A. A. Dameron, L. F. Charles, and P. S. Weiss, J. Am. Chem. Soc. **127**, 8697 (2005).

[112] A. A. Dameron, J. R. Hampton, R. K. Smith, T. J. Mullen, S. D. Gillmor, and P. S. Weiss, Nano Lett. **9**, 1834 (2005).

[113] S. Fujii, U. Akiba, and M. Fujihira, J. Am. Chem. Soc. **124**, 13629 (2002).

[114] T. Kitagawa, Y. Idomoto, H. Matsubara, D. Hobara, T. Kakiuchi, T. Okazaki, and K. Komatsu, J. Org. Chem. **71**, 1362 (2006).

[115] M. Kim, N. Hohmann, E. I. Morin, T. A. Daniel, and P. S. Weiss, J. Phys. Chem. A **113**, 3895 (2009).

[116] L. Landt, M. Staiger, D. Wolter, K. Klünder, P. Zimmermann, T. M. Willey, T. van Buuren, D. Brehmer, P. R. Schreiner, B. A. Tkachenko, A. A. Fokin, T. Möller, and C. Bostedt, J. Chem. Phys. **132**, 024710 (2010).

[117] D. Frost, F. Hering, A. Katrib, C. McDowell, and R. McLean, J. Phys. Chem. **76**, 1030 (1972).

[118] K. Ohno, K. Imal, S. Matmoto, and Y. Harada, J. Phys. Chem. **87**, 4346 (1983).

[119] V. P. Karpenko, J. H. Kinney, S. Kulkarni, K. Neufeld, and C. Poppe, Rev. Sci. Instr. **60**, 1451 (1989).

[120] L. Landt, C. Bostedt, D. Wolter, T. Möller, J. E. Dahl, R. M. K. Carlson, B. A. Tkachenko, A. A. Fokin, P. R. Schreiner, A. Kulesza, R. Mitrić, and V. Bonačić-Koutecký, J. Chem. Phys. **132**, 144305 (2010).

[121] A. A. Fokin, B. A. Tkachenko, P. A. Gunchenko, D. V. Gusev, and P. R. Schreiner, Chem. Eur. J. **11**, 7091 (2005).

[122] A. A. Fokin, P. A. Gunchenko, A. A. Novikovsky, T. E. Shubina, B. V. Chernyaev, J. E. P. Dahl, R. M. K. Carlson, A. G. Yurchenko, and P. R. Schreiner, Eur. J. Org. Chem. **30**, 5153 (2009).

[123] T. Willey, private communication.

[124] L. B. Clark and W. T. Simpson, J. Chem. Phys. **43**, 3666 (1965).

[125] A. Rauk and S. Collins, J. Mol. Spectr. **105**, 438 (1984).

[126] Y. Bing and P. Shou-Fu, Chin. Phys. B **17**, 1501 (2008).

[127] G. L. Vaghijani, J. Chem. Phys. **99**, 5936 (1993).

[128] Q. Shang and E. R. Bernstein, J. Chem. Phys. **100**, 8625 (1994).

[129] M. Hasan, D. Bethell, and M. Brust, J. Am. Chem. Soc. **124**, 1132 (2002).

[130] S. Letardi and F. Cleri, J. Chem. Phys. **120**, 10062 (2004).

[131] P. E. Pehrsson and T. W. Mercer, Surf. Sci. **460**, 49 (2000).

[132] S. J. Sque, R. Jones, and P. R. Briddon, Phys. Rev. B **73**, 085313 (2006).

[133] A. Gali, private communication.

[134] C. Kurtsiefer, S. Mayer, P. Zarda, and H. Weinfurter, Phys. Rev. Lett. **85**, 290 (2000).

[135] J. R. Maze, P. L. Stanwix, J. S. Hodges, S. Hong, J. M. Taylor, P. Cappellaro, L. Jiang, D. M. V. Gurudev E. Togan, A. S. Zibrov, A. Yacoby, R. L. Walsworth, and M. D. Lukin, Nature **455**, 644 (2008).

[136] G. Balasubramanian, I. Y. Chan, R. Kolesov, M. Al-Hmoud, J. Tisler, C. Shin, C. Kim, A. Wojcik, P. R. Hemmer, A. Krueger, T. Hanke, A. Leitensdorfer, R. Bratschitsch, F. Jelezko, and J. Wrachtrup, Nature **455**, 648 (2008).

[137] T. Möller *et al.*, Controlling the electronic structure of semiconductor nanoparticles by doping and hybrid formation, DFG research proposal.

Acknowledgements

The work presented in this thesis, which has been performed in the past three and a half years at the Technical University Berlin and in parts at the Lawrence Livermore and Lawrence Berkeley National Laboratories, has largely benefitted from the support of several people whom I would like to thank.

First I would like to thank Thomas Möller and Christoph Bostedt for supporting me in every possible way over the last years. I am very grateful to Thomas for giving me the opportunity to join his group and conduct the research that lead to this thesis. I am deeply indebted to Christoph for so many things that it will be difficult to list them all. Therefore I will only pick out his assistance with my research semester in Livermore, the guidance in scientific matters and beyond, and his friendship.

Trevor Willey not only facilitated my stay at the Lawrence Livermore National Laboratory, he also provided the XAS data of adamantane and adamantane-1-thiol in this thesis and a racing bike during my time in California which served me for several memorable rides up and down the Berkeley Hills. For all of the above and for proof-reading this thesis I am greatly thankful.

I had great time in the work group thanks to several group members without whom time in the office, the lab, and the *"thinking cabin"* would not have been half the fun. In particular I would like to thank Marcus Adolph, Tais Gorkhover, Matthias Hoener, Daniela Rupp, Sebastian Schorb, Matthias Staiger, Heiko Thomas, and David Wolter. I also gladly look back on a number of productive beamtimes and several entertaining night shifts with David Wolter, Witoslav Kielich, Matthias Staiger, Philipp Reiß, Stephanie Wutschik, and Bruno Langbehn. I probably still were not able to crawl more than one lane if it were not for Sebastian Schorb who was not only a great colleague and friend but also a fierce and pitiless training partner. The regular runs before and the swims after work provided the balance that was necessary to focus on the brainwork afterwards and the triathlons in Hamburg and Monterey were real highlights.[1]

This work would not have been possible without the contributions of Jeremy Dahl and Bob Carlson who provided the diamondoids and assured the purity of the

---

[1] *And you know I won, right?*

samples that was necessary for the present investigations. I am also thankful for the fruitful exchange of ideas. The second part of this thesis greatly benefitted from the work of Peter Schreiner and his group who produced numerous different modified diamondoids and kindly provided them to us for the present and several other studies.

I thankfully acknowledge the collaboration with the group of Prof. Arno Ehresmann from Kassel, and in particular Witoslav Kielich, Philipp Reiß, who provided crucial components and expertise for two joint photoluminescence beamtimes. Alexander Kulesza, Roland Mitrić and Prof. Vlasta Bonačić-Koutecký were the partners in another important and fruitful collaboration on the optical properties of diamondoid thiols. Their theoretical support substantially contributed to the outcome of this work. Further, I would like to thank Prof. Janina Maultzsch and Prof. Christian Thomsen for access to their Raman spectroscopy setup.

Finally, many thanks go to Robert Richter, Nils Rosenkranz, and Brandon Woolf for proof-reading chunks of this thesis.

Financial support from the Studienstiftung des Deutschen Volkes is gratefully acknowledged.

I want morebooks!

Buy your books fast and straightforward online - at one of world's fastest growing online book stores! Environmentally sound due to Print-on-Demand technologies.

Buy your books online at
www.morebooks.shop

Kaufen Sie Ihre Bücher schnell und unkompliziert online – auf einer der am schnellsten wachsenden Buchhandelsplattformen weltweit! Dank Print-On-Demand umwelt- und ressourcenschonend produziert.

Bücher schneller online kaufen
www.morebooks.shop

KS OmniScriptum Publishing
Brivibas gatve 197
LV-1039 Riga, Latvia
Telefax: +371 686 204 55

info@omniscriptum.com
www.omniscriptum.com

Printed by Books on Demand GmbH, Norderstedt / Germany